JOHN IKON
DVM, MSc., PhD, DIPL. ECVM

VETERINARY MEDICINE FOR ALL

THE BODY *of* DOMESTIC ANIMALS

Structure and function

PART 1

Veterinary Medicine for All.
The Body of Domestic Animals. Structure and Function

ISBN: 978-618-5627-13-3 (Ebook pdf, Part 1)
978-618-5627-12-6 (Ebook pdf, Part 1 and 2, set)
978-618-5627-16-4 (Ebook ePub, Part 1)
978-618-5627-14-0 (Ebook ePub, Part 1 and 2, set)
978-618-5627-04-1 (Print, Part 1)
978-618-5627-03-4 (Print, Parts 1 and 2, set)

First published edition 2021.

J.A. Ikonomopoulos
44 Thrasyboulou St. · Xalandri, Athens, Greece, 15234 · ikonomop@aua.gr ·

https://drhippocrates.com/en/

ABOUT THE AUTHOR

I WAS BORN IN 1967 IN ATHENS, GREECE, *and grew up in a small seaside village on the northern side of the Corinthian Bay, called Aspra Spitia (meaning "white houses").*

Even as a kid, I was fascinated by investigations and challenges! Step by step, these challenges have led me to the School of Veterinary Medicine of the Aristotle University of Thessaloniki, Greece, where I graduated in 1992. One year later, I was awarded the Master in Science degree in Veterinary Microbiology, by the Royal Veterinary College of the University of London. Just before completing my military service, I got the opportunity to use my postgraduate experience in molecular microbiology, for the diagnostic investigation of tuberculosis in humans. This was the beginning of my research career, and of my involvement in research project management, something that I gradually came to recognize as my "hidden talent"!

In 1996, I started my PhD project in the Medical School of Athens and the Veterinary School of Thessaloniki. Three years later I received my Doctor of Philosophy degree in Veterinary Science with Specialty in Veterinary Microbiology. The period that followed has been characterized by my inability to focus on one thing at a time; hence it is full of parallel activities, which resulted in many international collaborations and a number of awards, patents, and other distinctions. These activities have included practicing veterinary medicine, conducting post-doctoral research, coordinating international research projects, publishing research articles and books, providing diagnostic services to hospitals, and teaching.

Currently I am a professor in the Agricultural University of Athens, Greece since 2002, where I have had the pleasure to teach the Anatomy of Animals, Veterinary Microbiology, Animal Health, and Infectious Diseases of Animals, at both under and postgraduate level. A few years ago I had the opportunity to contribute to the foundation of the Hellenic Nanotechnology Society in Health Sciences, and to the European College of Veterinary Microbiology (ECVM). More recently, it was my great honor to be elected ECVM President and Rector of the School of Animal Sciences.

Today a father of two, I continue to be intrigued by challenges, one very fascinating of which was the book that you now hold in your hands.

Having realised with certainty his absolute intolerance to gratitude, I take great pleasure in thanking whole heartedly my good friend and talented page designer, Sotiris Papadimas, without whose help this book would have remained a cold assembly of texts and photos.
My gratitude is also due to my friend and good colleague in the European College of Veterinary Microbiology, Professor Bryan Markey of the University College Dublin for doing his best in turning my English into proper English!
Finally I must also thank my wife Athina for her love and support, my sons, Thomas for his brilliant suggestions and Aggelos for being very patient with me, and my many friends and colleagues who have contributed to this effort with their advice and guidance.

This book is dedicated to my beloved sons,
Aggelos and Thomas

A FEW WORDS FROM THE AUTHOR

I MUST CONFESS THAT MY INTENTION to write this book was delayed by a strong feeling of hesitation: who needs yet another book? how is the reader going to benefit from it, and what made me suitable to be its author?

Being in the pleasant position to combine three very inspiring properties, i.e. that of a father, a veterinary doctor, and a university teacher, I became convinced about the answer to the latter question. At the same time something that was becoming very obvious to me, especially in the last few years, was that veterinary medicine had started for some reason, to attract the attention and fascination of a gradually increasing number of people, of all ages! Soon I realised that although there are many excellent medical books for the general public, those available about veterinary medicine were either university textbooks or books for young children. However, my interest as a father was to find a book that would be **easy to follow but not oversimplified**. As a veterinary doctor, I always felt fascinated by the idea of a book that could demonstrate the beauty of my chosen profession, without hiding the fact that it demands **dedication and hard work**. Lastly, as a university teacher I was confronted since the early years of my career with the "ultimate challenge" that is universal to all those who have the privilege to teach: **how to make science inspiring!**

As you will have guessed, inspiration finally prevailed and this book was written with the **ambition** to become the first of a series that will cover all aspects of veterinary medicine. This series is primarily **addressed to** all those, whose interest in animal health could only be satisfied through an approach that is both accessible and scientific.

OUR JOURNEY INTO VETERINARY MEDICINE should undoubtedly begin with "Anatomy and Physiology of Animals", which in simple words is the description of the structure and function of the animals' body. Similarly, our guide on this journey could not be anyone other than Dr Hippocrates[1] who teaches and practices veterinary medicine. The name of our guide was a suggestion of my 10 year-old son but I must confess that I saw its value only when for the first time I paid attention to the literal meaning of the name of the famous ancient Greek physician.

info

1. Hippocrates, "Hippo" meaning "horse", and "crates" denoting "he who holds" or "tames"

Of course studying veterinary medicine with the help of Dr Hippocrates has its benefits, such as *"Stories from the life of Dr Hippocrates"*. These brief and sometimes amusing narrations refer to issues of common interest regarding animal health, such as neutering, vaccination, and disease transmission. The stories from the life of Dr Hippocrates are inspired by my own experiences and aim to make the reader understand that in veterinary practice the balancing of unexpected funny events, successes and of course failures, is something that every veterinarian is obliged to come to terms with, sooner or later.

Putting aside the literal meaning of the name Hippocrates, reference to the father of medicine is unavoidable, because of the importance of drawing attention to the history of Veterinary Science. Clearly, human and veterinary medicine went hand-to-hand for many centuries. Let's not forget that dissecting a human body, which is necessary for the study of human anatomy, was in many cases considered a sacrilege. In consequence, most of the names that were used to describe human bones derived from the study of the same bones in animals. Many of these terms that today seem peculiar and difficult to memorise, were once everyday words, whose original meaning has been lost over the centuries. Some of these words were first used in Ancient Greek whereas others were added in Latin, and together they became the basis of "modern anatomical terminology". However knowing the origin of these words demonstrates how close science really is to life. Hopefully, this "investigation" of terminology will inspire and encourage the reader, opening up a new world to **discoveries!**

Food for thought

Animals and Letters!

In the Phoenician language, the word "aleph" meant "ox". The word "aleph" gradually became "alpha" in Ancient Greek, and the horns of the ox rolled down to the side, to form the letter as we know it today.
The letter "a" is not the only one historically associated with animals. Continue this investigation with some research on the historical origin of the fourth letter of the alphabet!

Understandably, as more and more people around the globe come to reside in cities, many of those interested in veterinary medicine tend to associate pets with what is referred to as "domestic animal". As a result there is perhaps less interest into reading about the horse, sheep, goat, etc. However, approaching the anatomy and physiology of animals in a manner that includes **comparisons between all domestic animal species**, can help make us more

The history veterinary medicine at a glance

Name / When / Where
Urlugaledinna; 3000 B.C.; Mesopotamia
The **first man** ever recorded to have a passion for helping animals
Kahun papyrus; 1900 B.C.; Ancient Egypt
The **first written record** about Animal Health
Asoka; 1500 B.C.; Ancient India
Asoka (India's first Buddhist king) ensured that herbs for the treatment of animals would be available in his kingdom. Probably **the first case of administrative provisions focused on animal health.**
Horse-doctors; 500 B.C.; Ancient Greece
A class of physicians who provided treatment to horses; probably the **first group of professional** "veterinarians"
Lucius Junius Moderatus Columella; 1st century A.D.; Roman Empire
Prominent writer of books on animal care, health and breeding; Columella **used for the first time the term "veterinarious"**[1] to translate the Greek word for "horse- doctors" into Latin
Hippiatrica; 5th- 6th century A.D.; Eastern Roman Empire
A Byzantine compilation of Ancient Greek texts dedicated to the care and healing of horses; the main sources for this book probably come from the records of Hippocrates and other physicians of Ancient Greece.
Lord Mayor of London; 1356; London
Requests the specialists in equine hoof care (farriers) to form a fellowship for the improvement and regulation of how horses are treated
Worshipful Company of Farriers; 1674; London
The **first group of professionals** in modern times that are focused on improving animal care and health
Claude Bourgelat; 1761; Lyon, France
Founder of the **first veterinary school. The official beginning of** the veterinary profession.

familiar with those animals that are not part of our urban lives. Within this context, the analogies between the structure and the function of the human body with those of domestic animals are described for each anatomical system, especially with regards to the function of reproduction. Hopefully, in this way the reader will be able to comprehend more clearly the **"one health concept"**, or in other words that health protection has to be approached within the context of the interdependence of humans, animals and the environment.

info

1. *Vetus*, Latin for "old" or "experienced", as in "veteran" > *Veterinum* (Latin for "beast of burden")

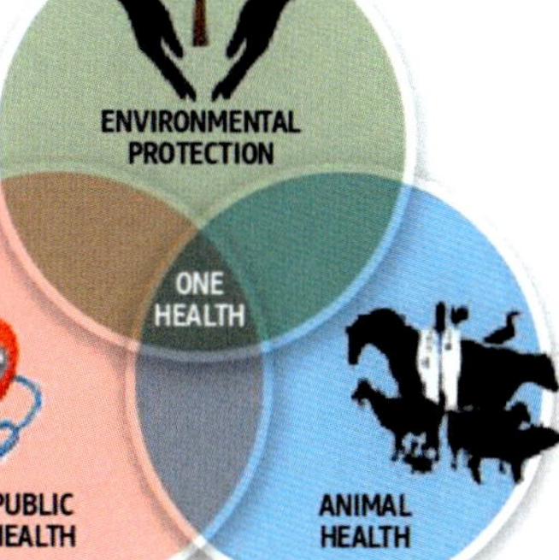

CONTENTS

1

COMPERATIVE FUNCTIONAL ANATOMY AND PHYSIOLOGY

The field of Biology that deals with the study of the body structure is called **Anatomy**[1] whereas **Physiology**[2] deals with the study of its function. Functional anatomy and physiology combines both topics by addressing body structure with reference to function. Using this framework it is much easier to understand why the body is structured the way it is, and how this helps survival and growth. The term "comparative" refers to the study of anatomy and physiology based on comparison between different animal species, including humans.

info

1. The author of the first textbook dedicated to anatomy was probably the famous physician of Ancient Greece, Hippocrates, who was born on the island of Kos, about 5 centuries B.C. It is likely that this book was not written by Hippocrates himself, but by his students. Hippocrates was probably the person who coined the term "anatomiki", meaning "to dissect" ("-tomy" the Greek for "section"). This term was adopted in different but similar forms into many languages: "*anatomia*" in Latin, "anatomie" in French and German, and in of course, "anatomy" in English.

Hippocrates of Kos

2. Physiology is a term that is used to describe a specific field of science, though in some cases it simply means "normal body function" (for example, in the next chapter you will read the expression "the physiology of the animal"). The term physiology is composed of two parts, "physio", which is the Greek for "nature" implying "normal" (as in **phys**ics i.e. the study of the laws of nature), and "logia", meaning "to speak about" (as in the words entomo**logy**, patho**logy**, epidemio**logy**, and many others).

SKELETAL AND MUSCULAR SYSTEMS

The **Skeletal System** consists of bones and joints, whereas the **Muscular System** consists of muscles; together they are often referred to as the Musculoskeletal System. These two systems that shape the body of humans and animals interact to generate body movement.

Skeletal System

The bones are rigid organs composed of minerals (mainly salts of calcium and phosphorus), organic substance (such as proteins and fatty acids), and water. The bones connect with each other at **joints**. These can be very mobile and are referred to **diarthroses** (e.g. the joints of the shoulder, the elbow or the wrist), slightly movable termed **amphiarthroses** (e.g. the joints between the vertebrae) or not movable at all, called **synarthroses** ⓘ (e.g. the joints between the bones of the pelvis or the skull).

info

1. The word "arthrosis" is the Greek for "joint". The word "diarthrosis" begins with the prefix "dia-", which usually indicates movement or connection, as in the words "**dia**rrhoea" and "**dia**meter". The word "amphiarthrosis" begins with the prefix "amphi-", which denotes "dual nature", like in the word "**amphi**bian", i.e. that which can survive both in water and on land. Finally, the word "synarthrosis" begins with "syn-", which indicates "addition", like for example the word "**syn**thesis".

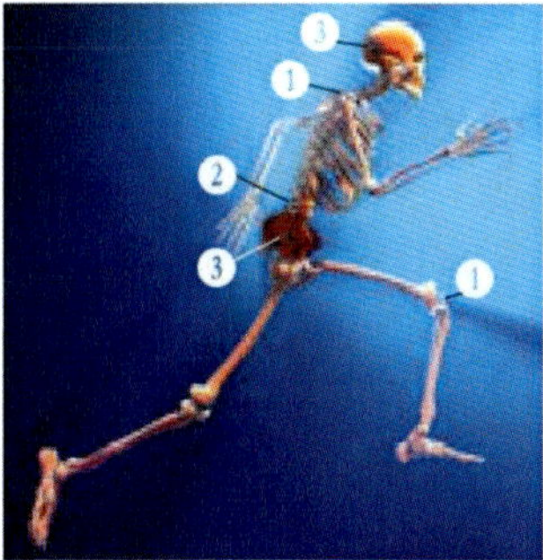

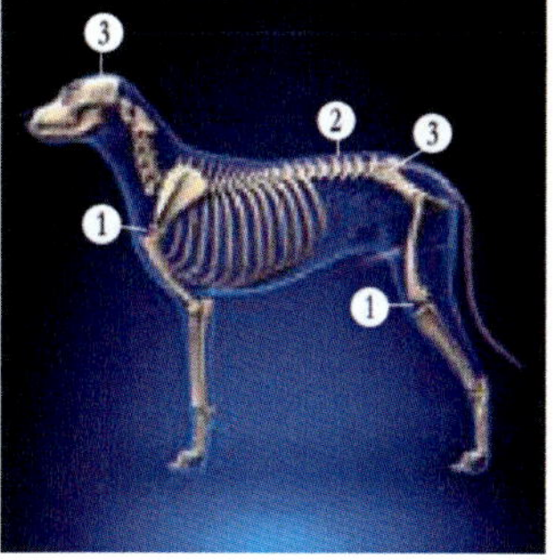

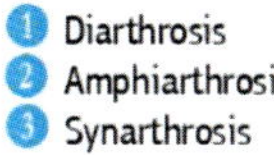
1 Diarthrosis
2 Amphiarthrosis
3 Synarthrosis

In addition to shaping the body, the skeleton has two very significant functions, to **protect** the sensitive organs of the body (including the brain, the heart, the stomach), and to act as **storage site for nutrients**, such as the salts of calcium that can be found in large amounts in the bones. When necessary, these nutrients can be mobilized from the skeleton into the blood, and thereby transported to all parts of the body.

Stories from the life of Dr Hippocrates

During pregnancy, the udder (the chest of the female) develops to be able to produce the milk. However, in order to provide the newborn with the calcium required to supply its needs, the specific mineral is mobilized within the body and directed into the milk with only a portion remaining to cover the needs of the mother. This means that if the mother is not taking the necessary amounts of calcium in her diet, she will become deficient, which can cause problems such as loss of teeth or even **spontaneous bone fractures**, i.e. fractures with no apparent cause.

Now that you know the basics about how calcium levels are managed in the body, it is perhaps more easy to understand why calcium shortage can be severe in cows that produce large amounts of milk. These animals may become unable to stand, lose consciousness or even be paralyzed. Luckily, this condition known as **milk fever** or **parturient hypocalcaemia**[1], can be treated easily with the administration of calcium to the sick mother.

When I was still a university student, one of my professors used to joke that cases of milk fever make the intervention of the veterinarian seem "divine", since with only one injection, a dying cow returns to normal, stands again on her feet, and "goes back to reading her newspaper!"

1. *Hypocalcaemia, "hypo-" is the Greek for "less" or "under" (as in "**hyp**othermia" meaning below normal body temperature), whereas "-calcaemia" means "**cal**cium in blood" ("**aem**a", is the Greek for "blood").

The standing position of animals

Standing on their toes!

If you compare the way humans and animals stand, you will realise that animals actually "walk on their toes". Can you think what purpose this serves? Would you believe that the standing position of animals can be the cause of serious health problems for them, or that it interferes with the way they sleep?

Animals stand on their digits, which increases their speed and ability to jump. Humans on the other hand, needed to maintain their upright position and therefore had to use a larger part of their leg for standing i.e. the region from the calcaneus (heel) to the digits. This reduced their speed but improved their stability and still requires a lot of practice in keeping body balance. This can be easily illustrated by observing newborn animals, which in most cases can stand and walk almost immediately after birth, whereas humans have to learn this through a process that usually takes more than 12 months.

This is of course important for those animals that must be able to walk or even run, as soon as possible, in order to evade danger.

Animal species	Age of starting to walk	Age of opening the eyes
Human	12-13 months	Immediately
Horse	30-60 minutes	Immediately
Dog	15-20 days	10-18 days
Cat	15-20 days	8-12 days
Lion	15 days	10-11 days

Another peculiarity of animals that is associated with their standing position and their need to be in constant alertness for predators is that **they can sleep on their feet!** Indeed horses, cows, elephants, giraffes, and many other large, four-legged land herbivores can "snooze on their hooves", and they lie down only when they want to sleep deeply.

However, do not think that the standing position of animals has no disadvantages. Tip toe walking and having to carry all of one's weight on the tips of the limbs, generates large pressure on the **tendons**, i.e. the structures that attach muscles to bones. Very often animals, especially race horses, suffer from **tendonitis**[1] (i.e. inflammation of the tendons), which creates an intense burning feeling in the affected area; the animals may limp and become irritated. Tendonitis is difficult to treat, and therefore is considered a serious condition, especially in high-value race horses.

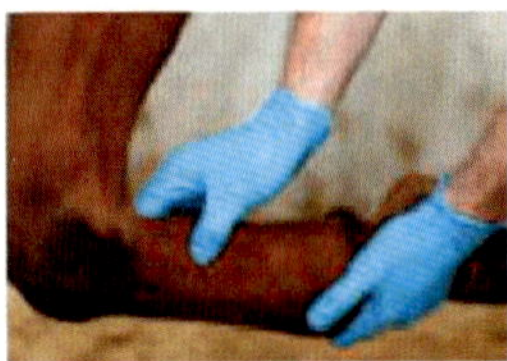

info

1. The **burning sensation** (i.e. "in flame") is one of the 4 basic characteristics of inflammation; the other three are **redness**, **swelling** and **pain**. These symptoms were first described by the Roman physician and writer Aulus Cornelius Celsus in the 1st century A.D. The body's normal response to **infection**, i.e. the invasion of pathogenic microorganisms, is usually inflammation. The term used for the inflammation of an organ is composed of the name of the organ and the suffix "itis". Hence "**otitis**" is the inflammation of the ear ("ota" the Ancient Greek for "ears"), and "**arthritis**", the inflammation of the joints ("arthrosis", the Greek for "joint").

The hidden message in the myth of Achilles' heel!

In Greek mythology, Achilles' mother, Thetis, dipped him in the water of the river Styx when he was a baby. The water of this magical river was supposed to make the baby invulnerable. However, to dip Achilles in the water, Thetis held him by the heel, leaving this as the only vulnerable part of his body. Eventually many years later, the great hero of the Trojan War is killed by a poisonous arrow shot right into his heel!

Isn't this myth a bit strange? Why should Thetis hold Achilles in the water of Styx by the heel, and not by his hand for example, which would be far more appropriate for a baby? Even so, how come she couldn't find a way to dip the heel too?

In those days, warriors covered their body parts with metal or thick layers of leather. However this was not as easy for that part of the leg that corresponds to the Achilles tendon, since this would have impeded their movement. The Achilles tendon is the largest tendon in the body and stretches from the heel to the calf muscles. When this tendon raptures it is very difficult to walk or even stand, and in ancient times, it would definitely have led to permanent disability. Therefore having your Achilles tendon injured in battle was not a small thing! At the same time, given that it was not usually protected by armour this part of the body represented a target for opponents trying to neutralize each other in combat.

Maybe it is exactly this fear for a vulnerable part of the body that was very strong among people back then, leading to all these strange details of Achilles' myth. One can't help wondering how appealing this myth would have been if Achilles were to survive the injury and to continue his life after the war, permanently disabled!

The spinal column

The **spinal column** (the spine) of animals is not much different from that of humans, and consists of multiple small units with a similar structure, the **vertebrae**[1]. Each vertebra is divided into two parts, the caudal, which is called vertebral body, and the dorsal, i.e. that part which is closer to the animal's back and is called the **vertebral arch**. As implied by its name, the inside of the vertebral arch is hollow and the vertebral arches of the vertebrae are aligned next to each other to form a tube that is called the **vertebral tube**. This tube provides protection to a very significant organ of the nervous system, the **spinal cord**. The nerves of the spinal cord pass through specific openings in the spinal column between the vertebrae and end up extending to the various muscles of the body.

1. Vertebral body
2. Vertebral arch
3. Spinal cord (inside the spinal column)
4. Intervertebral disk (nucleus)
5. Nerves of the spinal cord

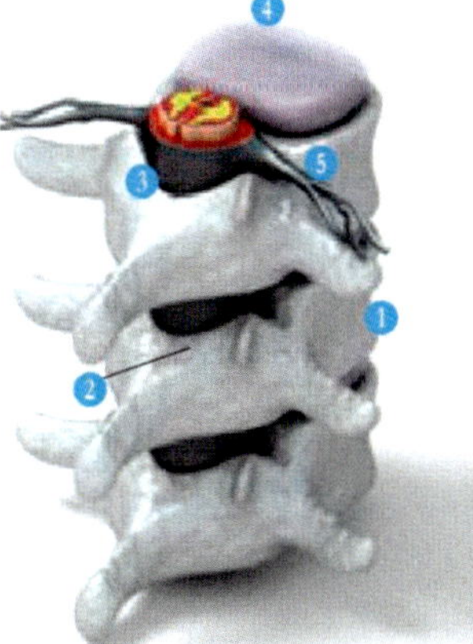

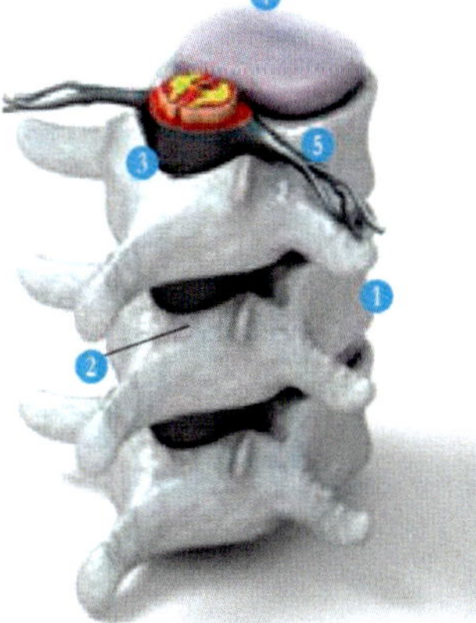

1. The plural form of "vertebra", from the Latin "*vertere*", meaning "to turn". These bones were given the name "vertebra", probably because they were considered responsible for the body's ability to turn.

The body changes with age

The body consists of bone, muscle and fat, but the relative amount of these tissues changes with age; for example in the pig:

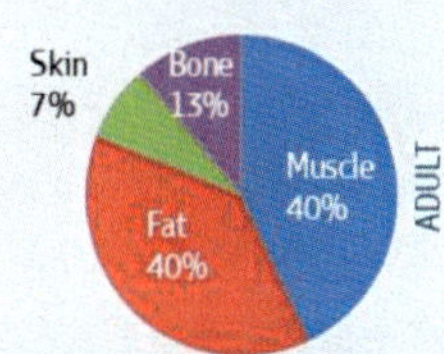

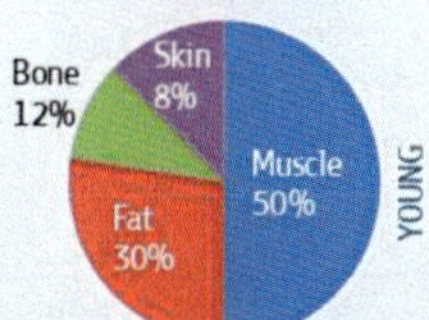

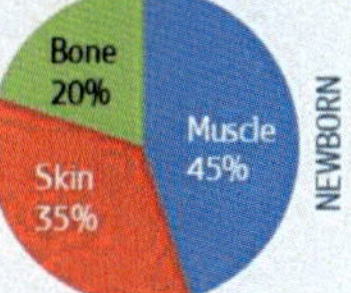

The bones of the Skeletal System

The bones of the front limb
(in humans these are called upper limbs)

1. Scapula[1]
2. Humerus
3. Antebranchium (consisted of radius and ulna)
4. Carpus (wrist)
5. Metacarpus[2]
6. Digits

The bones of the hind limb
(in humans these are called lower limbs)

7. Pelvis
8. Femur
9. Crus (consisted of tibia[3] and fibula[4])
10. Tarsus (hock) consisting of talus (ankle) and calcaneus (heel)
11. Metatarsus[5]
12. Digits

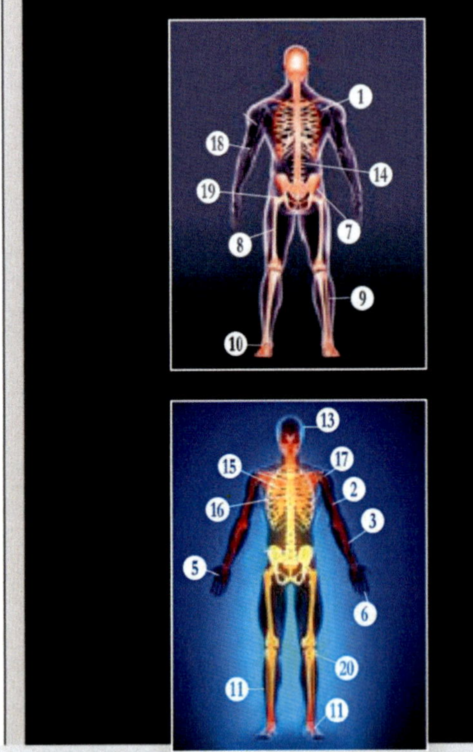

1. Scapula, Latin word meaning "shoulder" or "shoulder blade". Maybe the original meaning of the word was "shovels"; indeed the specific bone resembles a small shovel.
2. Metacarpus, that which is "meta" (Greek for "after") the "carpus", i.e. the wrist. The word "carpus" probably comes from the word "carfos" which was used by the ancient Greek epic poet Homer with the meaning "dry area", i.e. area with no vegetation. The term was probably used in anatomy as a metaphor, since this part of the body has no meat (muscle).
3. Tibia, Latin for "pipe, flute", indeed this bone is long and straight like a flute
4. Fibula, Latin for "pin/clasp". The term used in Greek for the specific bone ("oleni") has the same meaning, obviously because of its resemblance to a pin.
5. Metatarsus, exactly like the word "carpus" (wrist), the word "tarsus" implies that this body part has no meat

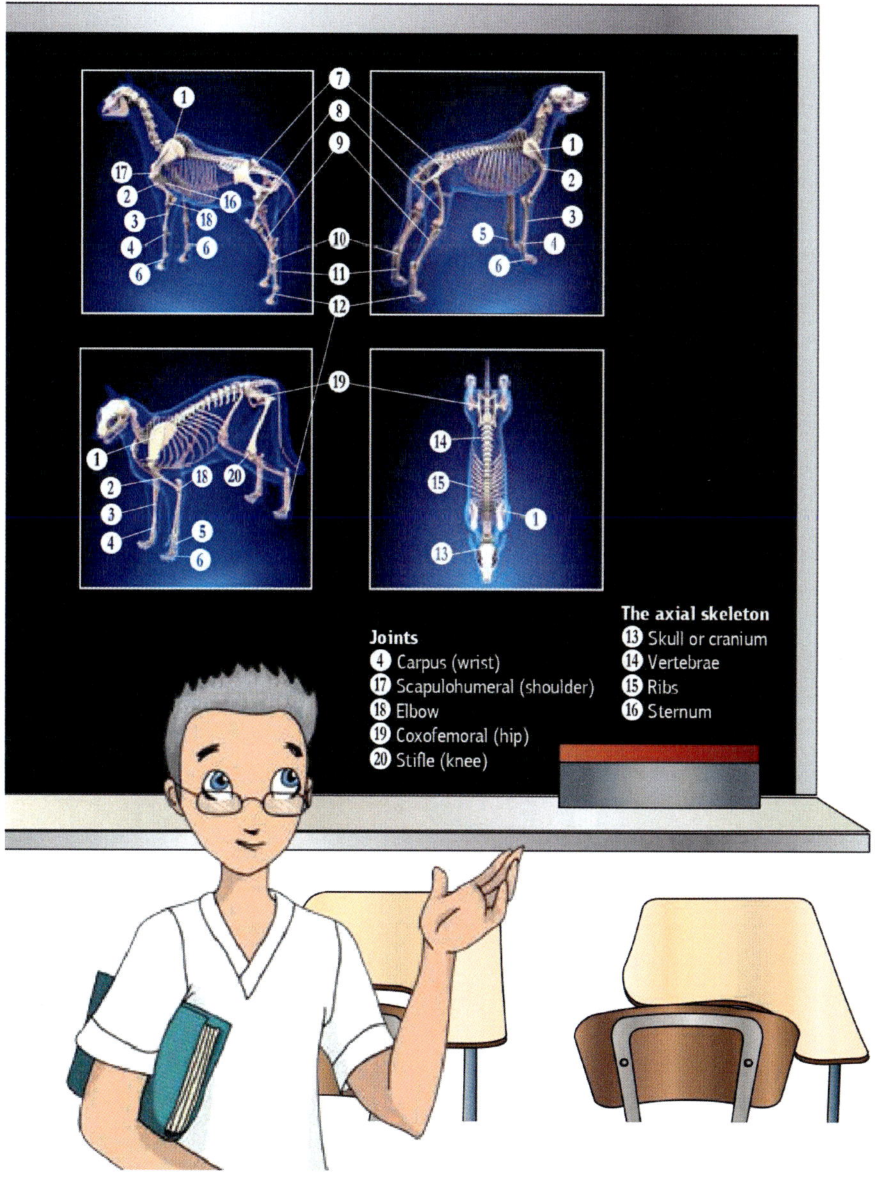
Joints
4 Carpus (wrist)
17 Scapulohumeral (shoulder)
18 Elbow
19 Coxofemoral (hip)
20 Stifle (knee)
The axial skeleton
13 Skull or cranium
14 Vertebrae
15 Ribs
16 Sternum

At this point it would be a good idea to study the structure of the spinal column and investigate how this might be associated with disease:

*The gelatinous nucleus of the intervertebral disk is dislocated (**prolapsed intervertebral disk**) and pressing on the adjacent nerve of the spinal cord*

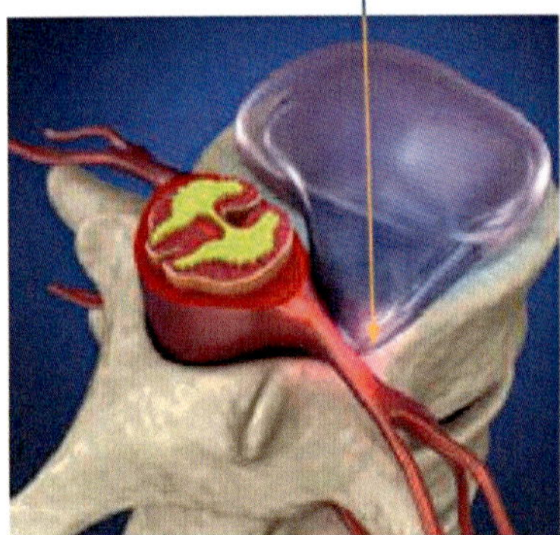

Adjacent vertebrae are not directly apposed to each other; instead soft structures called **intervertebral discs** are positioned between adjacent vertebrae. The central part of these disks consists of a gel-like (gelatinous) substance called the nucleus (**nucleus pulposus**). The nucleus functions very much like a cushion or a pillow that absorbs part of the large pressures that are applied to the spinal column during locomotion i.e. walking, running or jumping. This nucleus may become displaced out of its normal position (dislocated), because it is held in place by cartilage (**annulus fibrosus**), which breaks rather easily. In any other part of the body this dislocation would not be a great problem, but in the spinal column it can be very serious! This is because the nucleus pulposus is located very close to the spinal cord and its dislocation generates pressure on the adjacent nerves, causing intense pain and sometimes paralysis. This problem is quite common in veterinary practice and usually results from mechanical injury caused for example by fights between animals, intense exercise (e.g. horse-racing) or car accidents.

THE SKULL IS POSITIONED UPON the first of the vertebrae which is named **atlas**[1],after the titan who held the sky on his shoulders. There are several differences in the shape of atlas compared with that of the other vertebrae, the most significant being that atlas has no vertebral body. Being ring-shaped,

info

1. In Greek mythology, the mightiest and most skillful titan, Atlas, was their leader in the battle against the gods of Olympus, i.e. the Titanomachy. The battle was victorious for gods and their leader Zeus, condemned Atlas to hold the weight of the sky forever from the top of a high mountain in Africa. The name "Atlas" implies the great endurance of the titan who suffers this endless torture as his punishment ("a-tlan" = "being able to endure anything").

atlas provides a stable support to the skull and can rotate around the straight line that passes through the 2nd vertebra, which is aptly named **axis**.

The base of the skull (**occipital bone**[1]), forms, together with atlas and axis, two joints that are specifically constructed to allow greater movement than all the other vertebral joints (i.e. the joints between vertebrae). As a result, the way the atlas is attached to the occipital bone (**atlanto-occipital joint**) allows the neck flexion, i.e. to move the head as if to say "yes". The connection between the atlas and the axis (**atlanto-axial joint**), allows head rotation, i.e. the ability to shake the head as if to say "no".

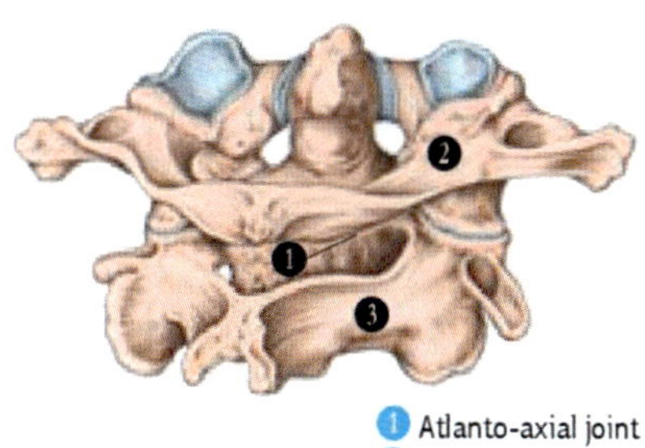

1 Atlanto-axial joint
2 Atlas
3 Axis

1 Occipital bone
2 Atlanto-occipital joint
3 Atlanto-axial joint

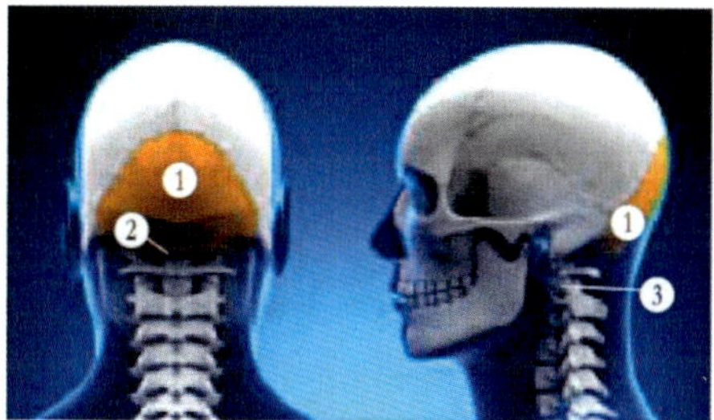

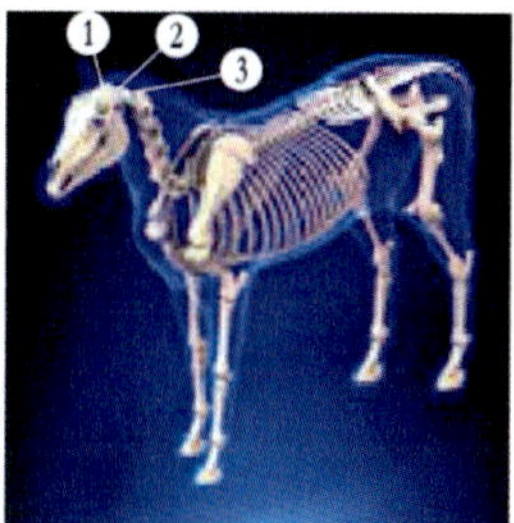

THE SKELETAL SYSTEM OF MANY animals evolved to allow far greater movement of the head compared with humans.

This is also true for birds that do not move their eyes very much, so their skeleton is designed to allow very rapid and broad rotation of the head. Of course, the champion of head-rotation is none other than the owl that seems to be able to turn its head almost a full circle!

This owl is actually using only half of its head rotation capacity

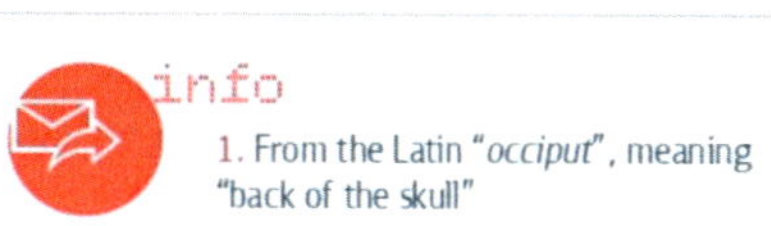

1. From the Latin "*occiput*", meaning "back of the skull"

Maybe this cute animal reminds you of something: the tarsier was the inspiration behind "Master Yoda", the famous character of the movie Star Wars!

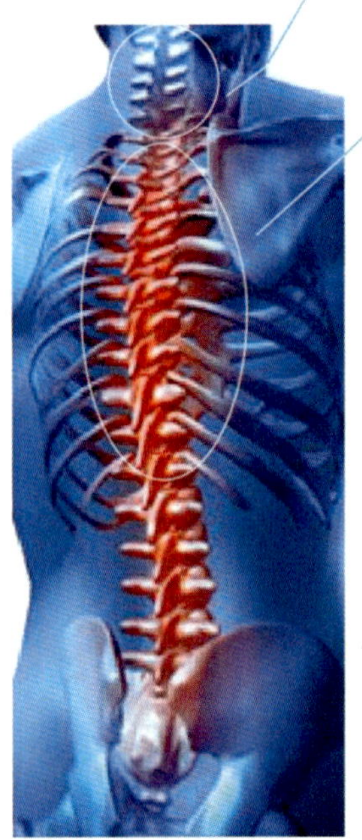

The mammal with the greatest head rotation ability is a small primate, native to Southeast Asia, called a **tarsier**. Tarsiers love to eat insects and they are great at catching them, which is of course facilitated by their ability to turn their head nearly 180° to each direction.

THE SPINAL COLUMN of humans and animals has similar structure and can be divided into the same **5 regions**:

- The **cervical**[1] **region** of the spinal column corresponds to the "neck". The cervical region consists of 7 vertebrae in most mammals (including the giraffe, despite its extremely long neck), the first two of which are atlas and axis. In the chicken, this region consists of twice as many vertebrae (14), and is shaped like an "S", which improves considerably body balance, especially when it stands on its feet.
- The **thoracic region** of the spinal column corresponds to the chest, and in most domestic animals consists of 13 vertebrae, in the human 12, in the horse 8, but in the chicken only 7! A special characteristic of the thoracic vertebrae is that they support the ribs, to which they are anchored by joints. In the front wall of the chest the ribs are attached to the sternum[2]. This arrangement, i.e. thoracic vertebrae – ribs – sternum, results in the formation of the **thorax**, a cage like structure that surrounds a cavity called the **thoracic cavity**, which contains

info 

1. Cervix, a word with many uses! The word "cervical" means that which belongs to "*cervix*", which is the Latin for "neck". The first part of the word "*cervix*", i.e. "cer-", comes from the Ancient Greek word "keras", meaning "horn". In animals, the uterus presents two structures that look like horns, and the part of uterus that is located just before these "horns" is called "cervix", exactly like the part of the body that is located just before the head carrying the horns, which is also called "cervix".

2. From the Ancient Greek word "sternon", the original meaning of which was "broad/extended". It seems that in those days, the word "sternon" was used to describe not the specific bone, but the entire front part of the chest that is indeed broad. This word was suitable for use only in connection with males; the respective word for the female was "stethos", like in the word "**stetho**scope". It is perhaps interesting to note that the stethoscope was devised and first used to listen to the sounds of the lungs only in women, since in men it was not considered inappropriate for the doctor to put his ear against their chest or back.

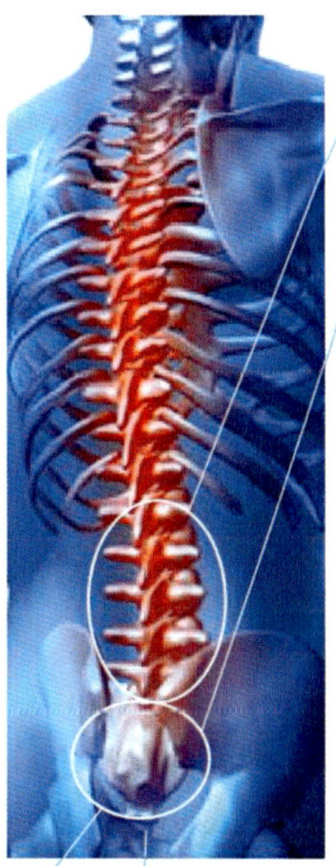

Sacral and coccyx region

the heart and lungs. This arrangement provides protection for these vital organs.

The **lumbar**[1] **region** of the spinal column corresponds to the middle part of the animal's back i.e. that part between the thoracic region and the pelvis. The lumbar region consists of 5 vertebrae in humans and slightly more (6-7) in all domestic animals.

The **sacral**[2] **region** of the spinal column corresponds to the part of the body that is cranially to (above) the hind limbs. In all domestic animals and humans, the sacral region consists of 4-5 vertebrae (sacral vertebrae) that, around adulthood, fuse into a single bone, the **sacrum**. In the chicken, the sacral vertebrae are fused together with the lumbar vertebrae forming a single bone consisting of 14 vertebrae. This provides stronger support to the wings during flight and improves their aerodynamic thrust.

The pelvis is located right underneath the sacrum and consists of two bones, the **anonymous**[3] bones, each consisting of three smaller bones: the **ilium**[4], the **ischium**[5] and the **pubis**[6]. The two anonymous bones are connected together in the front (humans) or the ventral part (animals) of the pelvis, forming the **symphysis**[7] **pelvis**. This arrangement, i.e. sacrum – anonymous bones – symphysis pelvis, results in the formation of another cavity, the **pelvic cavity**, which protects the uterus of the female and the urinary bladder.

info

1. Lumbar, from the Latin "*lumbus*", which means "loins"
2. Sacral, from the Latin "*sacrum*", meaning "sacred". The name of this bone is perhaps an example that things did not always work out great for the scholars of anatomy: the specific bone was named by the Greek physician Galen (129-200 A.D.) with the word "iero". In those days, the word "iero" had two distinct meanings in Greek, "sacred" and "powerful", but the latter was not very common (in Modern Greek, the word "iero" is used to denote only that which is sacred). Unluckily, those who used the term "sacrum" for the specific bone actually translated into Latin, the meaning of the word that was more common, and not the one denoting its specific physical characteristic i.e. being strong.
3. Anonymous means that which has no name. It seems as if the pioneers of anatomy ran out of imagination for these particular bones, which for some reason were left without a name; hence they are called "anonymous".
4. Ilium, singular created from the Latin word "*ilia*", meaning "groin"
5. Ischium, from the Greek word "iskhion", meaning "the hip"
6. Pubis, from the Latin word "*pubes*" (also the basis of the word "puberty"), meaning "genital area", or "groin"
7. Symphysis, from the Greek word "physis", meaning "growth"; hence the word "sym**physis**" meaning "that which is growing together".

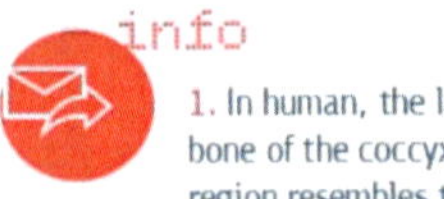

1. In human, the last bone of the coccyx region resembles the cuckoo's beak; "coccyx" means that which looks like the cuckoo. The name of the bird is echoic of its cry.

The **coccyx**[1] **region** (tailbone) of the spinal column corresponds to the "tail". In the human, coccyx consists of 4 small vertebrae, whereas in the chicken it consists of 6, and in most domestic animals of 16-23 vertebrae (20-23 in the dog and the cat; 16-18 in the sheep; 12 in the goat, and 15-20 in the horse). In humans and other tailless primates such as the great ape, the vertebrae of coccyx are fused together. In many animal species however, these bones are separate, and this transforms the tail into a very useful tool of balance and protection against insects.

In one class of hunting dog breeds, called the "pointing breeds", the tail has one more, rather peculiar, use: it is stretched horizontally forming a straight line (an arrow) with the dog's head and body, indicating the location of game.

In addition to the dog, many other animal species use their tail in a way that indicates clearly their mood and intentions.

1. Cervical region
2. Thoracic region
3. Lumbar region
4. Sacral region
5. Coccyx region

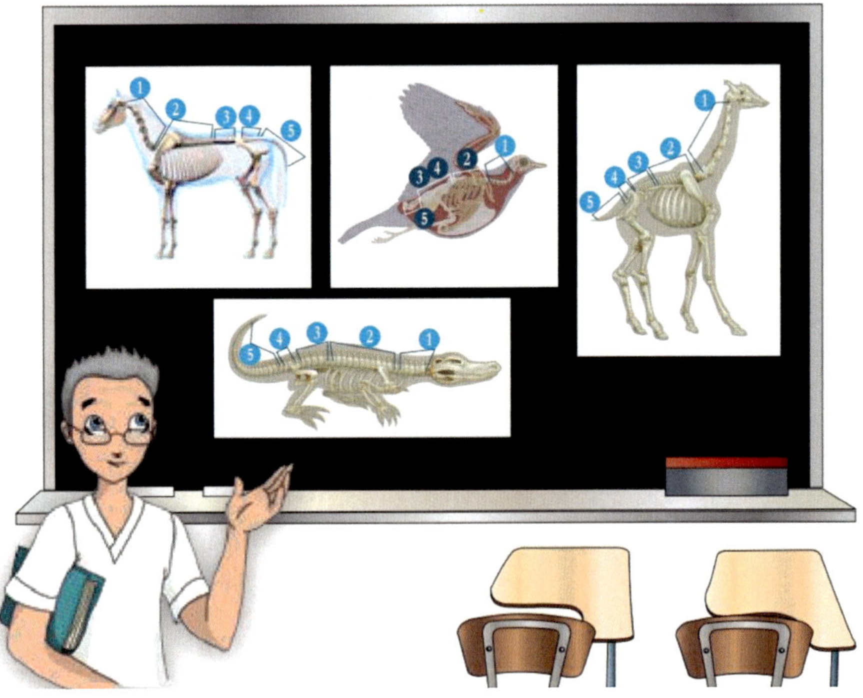

Of course there are also some dog breeds, for example Akitas, a breed of dogs originating from Japan, whose tail bones are anchored together to a twisting wedge shape. In this breeds the tail is coiled above or behind their back.

Tail feathers to show off!

The position and the movement of the tail	Mood/Intention
Upwards and wagging	Playful, "I'm feeling happy"
Underneath the hind limbs	Scared, "I need help"
Stretched to the back	Alertness, "something caught my attention and I need to react immediately"

Muscular System

In addition to the movements of the skeleton, the functions of the Muscular System include transportation of ingested food through the stomach and the intestine, generation of body heat, circulation of blood and regulation of its flow to body parts. The Muscular System represents approximately half of body weight, which gives an indication of its diverse functions and the amount of work required to perform them.

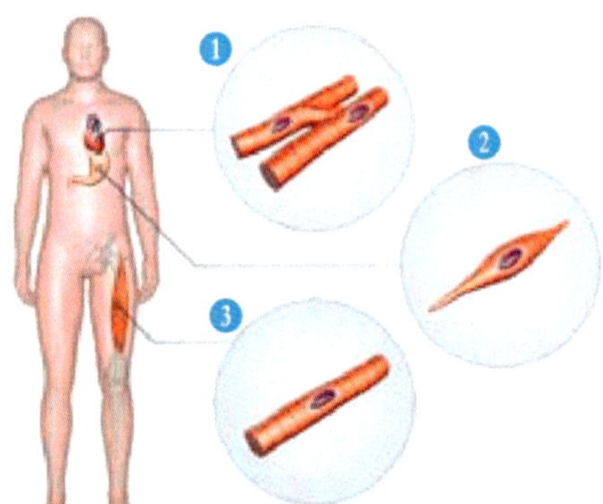

1 Cardiac muscle (striated)
2 Smooth muscle
3 Skeletal muscle (striated)

The muscles consist of **muscle cells**. There are three types of muscle cells in the animal body: **skeletal**, **smooth**, and **cardiac**.

The skeletal muscle cells, also called **muscle fibers**ⓘ, can be divided into three types, based on their microscopic appearance: the **red**, the **pale** (or white), and **intermediate muscle cells**. In general, muscles consist of a mixture of these three types of cells, but depending on animal species, a muscle may consist of more red or white muscle fibers. For example, the red muscle cells predominate in the breast muscle of pigeons, but the oposite occurs in the chicken. This makes pigeons able to sustain flight, just like geese and ducks, because red muscle cells **contract** (become shorter) more slowly compared to white, but fatigue less rapidly. When viewed under the microscope, skeletal muscles appear as if they are divided by vertical lines, which gives them a striated (striped) appearance. Skeletal muscles are attached to bones, and are responsible for **voluntary movement**, i.e. the movement of body parts, which is controlled by will.

info

1. If you observe their shape under the microscope, you will notice that only skeletal muscle cells are thin and long, i.e. only this type of muscle cells appear like "fibers". Therefore the term "muscle fibers" is correct only in connection with the skeletal muscle cells. This observation was first made by Anton van Leewenhoek, a Dutch merchant of cloth, who is recognized as the "father of microscopy". Van Leewenhoek died in 1723 at the age of 90, with a very significant record of scientific discoveries, including the existence of microbes that he named "animalcules", meaning "tiny animals".

Van Leewenhoek's "animalcules" ingeniously drawn!

The smooth muscle cells are so named because they have no visible striation. This type of muscle cells make up part of the internal organs such as the stomach, intestine and lungs, and their function is not controlled by will **(involuntary movement)**.

The cardiac muscle cell is found only in the heart and, though striated like skeletal muscle cells, its function is not controlled by will. The cardiac muscle of humans and animals forms the wall of the heart, and it is called myocardium[1].

Regardless of type, muscle cells, and effectively muscles, have a similar mode of action. They have two ends that are attached to different parts of the body, such as between bones that form joints (skeletal muscles) or specific parts of internal organs like the intestine (smooth muscles). In some skeletal muscles, one of their ends spreads in 2 or more sections, with which they attach to different sides of a joint or a bone, such as for example the biceps[2] or the triceps muscle. This arrangement makes the attachment of the muscle to the bone stronger and the movement of large joints smoother.

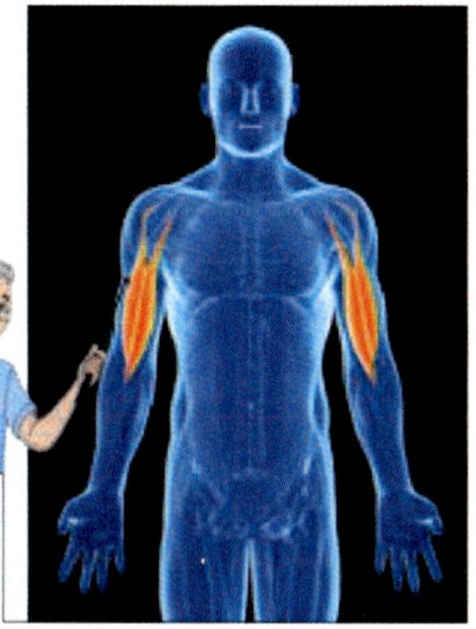

The "two heads" of the biceps (biceps brachii) of the human

When the muscle cells contract, they become shorter and this generates either **movement** or **resistance**. An example of muscle contraction generating movement is that of the skeletal muscles that move a body part. An example of muscle contraction generating resistance is that of the muscles located inside the wall of blood vessels. When these muscles contract, the width of the blood vessel decreases, and this causes a decrease in blood flow.

1 The masseter, the strongest of the masticators

info

1. Myocardium is the muscle ("mys", the Ancient Greek for "muscle" and "mouse") of the heart ("cardia", the Greek for "heart"). The two meanings of the word "mys" relate obviously to the resemblance of the shape of some muscles to mice. The Greek "mys" became "*mus*" in Latin, and this was the origin of the words "muscle" and "mouse".

2. The first part of the terms, i.e. "bi" and "tri" denotes number ("2" and "3"), whereas the second part, "ceps", comes from the Latin word "*caput*", meaning "head"; hence "biceps" denotes "two-headed".

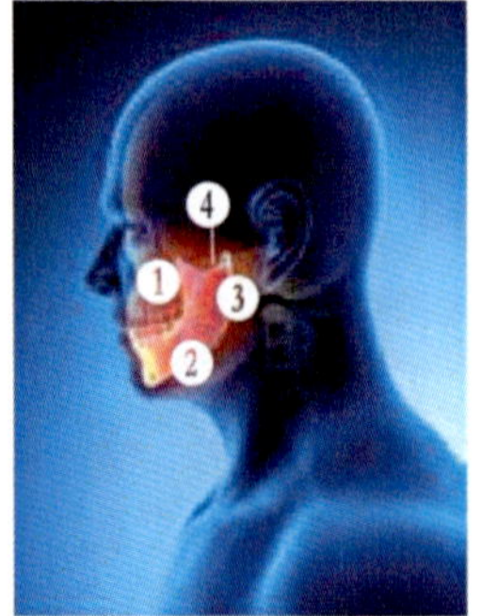

1 Maxilla
2 Mandible (central part)
3 Mandible (vertical part)
4 Temporomandibular joint

THE ABILITY OF SKELETAL muscles to generate movement is improved by having one end attached to a part of the skeleton (**origin of the muscle**) that is less movable than the other (**insertion of the muscle**). This means that when a muscle contracts, its energy is applied on both ends but this generates movement mainly at the point of insertion.

Let's try to understand how muscles work, using the process of **mastication** (chewing) as example:

Mastication involves basically 3 bones and 5 muscles, on each side of the head. One of these bones is the lower jaw i.e. the **mandible**[1], the other is the upper jaw i.e. the **maxilla**[2], and the third is the **temporal**[3] **bone**, which corresponds to the sides of the head, known as the temples. The mandible is the only movable part of the skull; it is U-shaped and consists of a central part, which forms the chin and supports the lower teeth, and two vertical parts on each side of the skull that connect (**articulate**) with the temporal bone, through a very flexible joint called the **temporomandibular joint**.

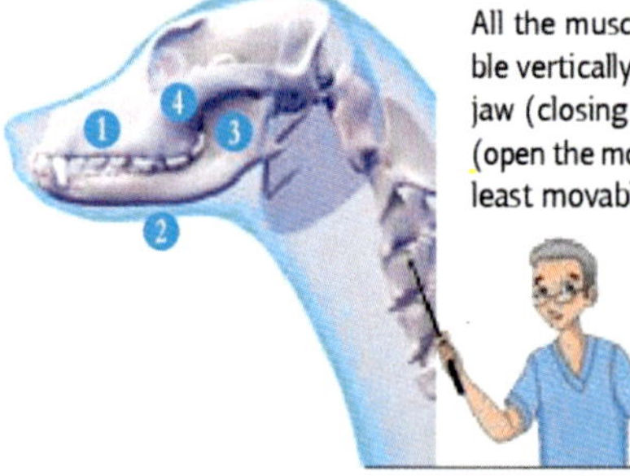

All the muscles of mastication (masticators) move the mandible vertically; four of them are responsible for **adduction** of the jaw (closing the mouth), and only one is required to **abduct** it (open the mouth). This means that the origin of masticators (the least movable part) is the temporal bone and the maxilla, and their insertion (the more movable part) is the mandible. When the adductors contract, their energy is applied at both sides of the mouth (temporal bone and maxilla above, and mandible below), but the only part that moves is of course the mandible.

Mouth abduction

info

1. Mandible, from the Latin "*mandibula*", meaning "jaw". 2. Maxilla, from the Latin "*maxilla*", meaning the upper jaw. 3. Temporal, from the Latin "*tempus*", meaning "time" or "season" (also the origin of the word "temporary"). The name of the bone is propably associated with the fact that the specific part of the head presents signs of how "time passes": the hair growing grey at the temples, and the pulsations of the artery located above the bone (temporal artery) indicating the heartbeat.

The temporomandibular joint is one of the most complicated joints of the body, and performs a great variety of movements, i.e. up and down, backwards-forwards, and side to side, in various combinations. This is possible because the 5 muscles mentioned above, and the smaller muscles that support them can move both independently and in combination with each other.
In humans, these moves are also necessary for communication (talking, laughing etc), but in animals they are mainly associated with mastication. The flexibility of this joint is more significant in herbivores like the horse, cow, sheep and goat, because in addition to cutting their food in small pieces, herbivores have to grind it between their teeth and mix it thoroughly with saliva, which makes it easier to digest. In these animals chewing takes several hours per day, and therefore requires masticators that can cope with a heavy workload.

It is probably easy to understand why it takes four muscles to close the mouth, and only one to open it, since the ability to chew is defined only by the strength with which animals can close their mouth. The ability to bite forcefully is very important in animals because they depend on the strength of their masticators not only for chewing but also for survival. Recently it was discovered that bite force does not depend only on the size of the animal, but also on the size of its prey.

info

Strongest bite

Ranking order *(from weaker to stronger)*	Animal species	Bite force *(compared to bear)*
1	Bear	1
2	Tiger	1.1
3	Hyaena	1.2
4	Gorilla, Shark	1.3
5	Jaguar	2
6	Alligator	2.3
7	Crocodile	4
8	Nile crocodile	5

Measuring the bite force of animals leads to the conclusion that the champion of biting force is by far the Nile crocodile, who bites almost 3 times stronger than the shark, and 5 times as strong as the bear!

Nile crocodile

Anaconda

American eagle

IN SOME ANIMAL SPECIES, the Muscular System acquires particular significance for their survival in nature:

- Anaconda: There are four types of anacondas in the world, but the biggest is the green anaconda, which weighs up to 220Kg! All anacondas are non-venomous snakes and kill their prey using their ability to constrict, i.e. to squeeze their victims[1]. This requires enormous strength, which means that for anacondas to survive in nature they need a particularly well-developed muscular system.
- Eagle: In terms of their ability to lift weights, eagles are definitely the strongest birds in the world. Very often, eagles will use their powerful feet and wings to lift off their prey, which can be up to 7 times their own weight!
- Ant: Many kinds of ants collect leaves that can be up to 50 times heavier than their body. The great strength of ants is generated from a muscular system that is much more developed compared to even the strongest mammal. Furthermore, to be able to cut leaves quickly, ants possess another peculiar ability, which is also associated with their highly developed muscular system. They can vibrate their jaws at the unbelievable speed of 1,000 times per second, which makes them capable of grinding large leaves in seconds!

Two of the strongest creatures on earth!

There are of course many other very strong animals, like the bear, the elephant and the ox, but the **strongest of them all is by far the dung beetle**. The adult male dung, not more than a few centimeters long, can lift weights about 1,100 times its own, which corresponds to a man being able to lift 5 large buses! Of course this requires not only an incredibly developed muscular system, but also a skeletal system that would not be crushed by this enormous weight, which is why these beetles have a very strong **exoskeleton**[2] and hard forewings.

info

1. For many years, scientists debated about what happens to victims of constriction; some claimed that it breaks their bones, and others that it results in suffocatione. Recently it was discovered that constriction leads to the death of an animal because of **brain ischemia** i.e. a lack of blood flow to the brain, which is caused by the pressure applied on blood vessels. This condition leads first to a loss of consciousness, and then slowly to death. 2. Exoskeleton: unlike the skeleton of mammals which is inside their body (**endoskeleton**), exoskeleton covers the body externally, like armor, providing support and protection.

The Muscular System is arranged in a similar manner in humans and animals, and in several cases their muscles have identical names, like for example those referred to below:

1. Biceps branchii
2. Pectoralis major
3. Quadriceps femoris
4. Deltoid[1] muscle
5. Trapezius
6. Triceps branchii
7. Santorius
8. Gastrocnemius
9. Gluteal muscles
10. Achilles tendon

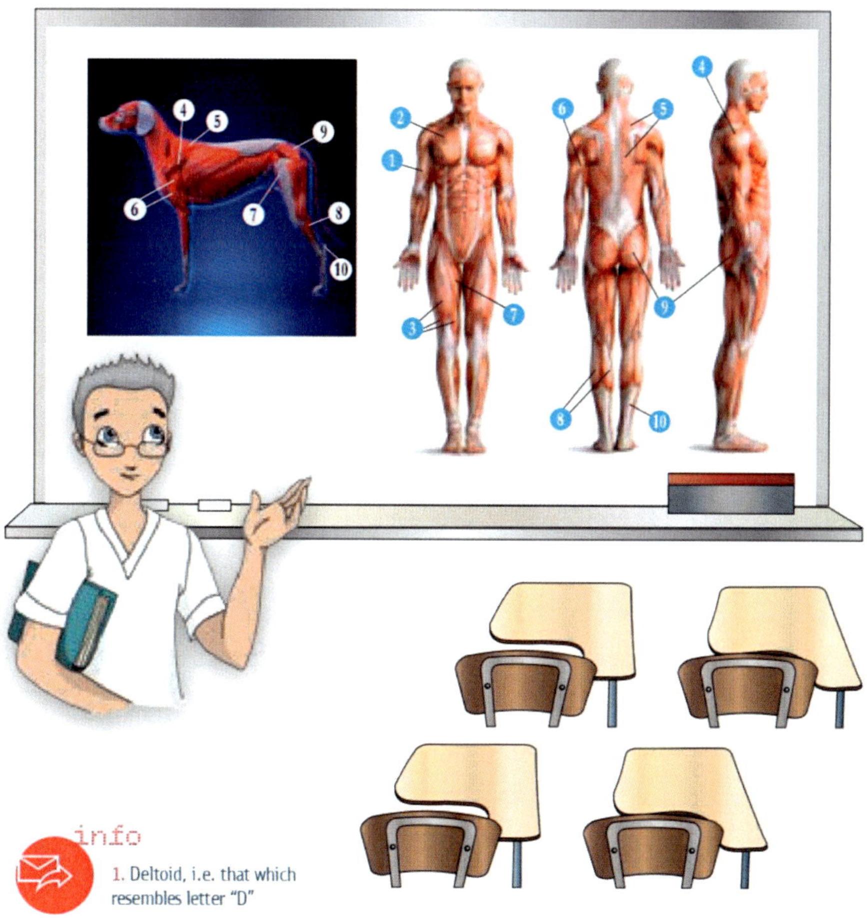

info

1. Deltoid, i.e. that which resembles letter "D"

NERVOUS SYSTEM

The Nervous System is an extremely advanced coordination unit, which is of great importance in maintaining normal body functions and health.

As in humans, the primary function of the Nervous System of animals is to receive input (**stimuli**) from both inside the body and from the environment. These stimuli are then processed in specific regions that are called **integration areas**, to formulate the animal's self-consciousness and perception of its surrounding (**sensory function**), and to coordinate accordingly the activity of its muscles and organs (**motor function**).

The interaction between sensory and motor functions with integration is responsible for the animal's ability to learn, generate emotions and respond to the changing conditions of the environment. Effectively, the Nervous System defines what is referred to as "character", "talent" and "intelligence". In humans we know very well that these properties differ from one person to the other. This is also true not only for different animal species but also for different individual animals. A typical example of this is the cat: there are some very calm and cuddly, while others are very energetic and independent.

Before we discuss in more detail the characteristics of the Nervous System, let's study first its basic structure and functions.

The Nervous System has an extremely complicated structure but is organised in a rather simple manner, which makes it very effective and efficient. It would not be inaccurate to state that the Nervous System is organised very much like a computer, i.e. there is a central processing unit which functions as an "administrative centre" and a peripheral system. The "central unit" of the Nervous System is referred to as the **central nervous system** and consists of the **brain** and the **spinal cord**.

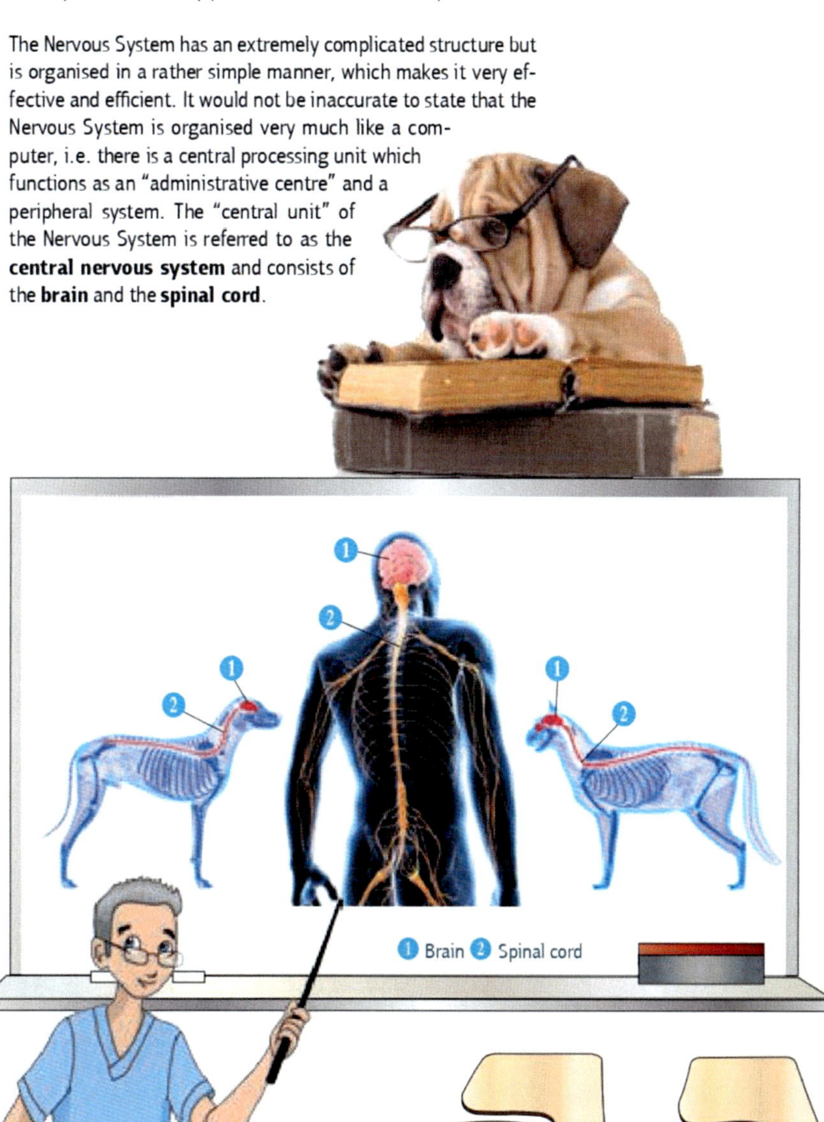

The "peripheral system" of a computer would correspond to the **peripheral nervous system**, which consists of **nerves** and sensory structures that can be highly specialized organs like the eye, or simpler structures that are called **receptors**. The sensory structures and the nerves of the peripheral nervous system correspond to the screen, the mouse, the keyboard and the wires of a computer.

The brain is the central processing unit of the very advanced "computer" that controls the body

The central nervous system is of vital significance but also very vulnerable to damage. Therefore the brain and the spinal cord are wrapped in three layers of thin membrane, which are called **meninges**[1], and they are located in very well protected "containers" made of bone i.e. the cranium that surrounds the brain, and the vertebral column which contains the spinal cord. For more protection and nourishment, the space between the meninges and the brain or the spinal cord is filled with a liquid called **cerebrospinal fluid.**

1. Skull
2. Dura mater, (*"dura"* Latin for "strong", "mater" the Greek for "mother") the strongest of the meninges
3. Arachnoid mater, resembling spider web ("arachnid" the Greek for "spider")
4. Pia mater, the most delicate (*pia*) of the meninges
5. Brain
6. Spinal cord

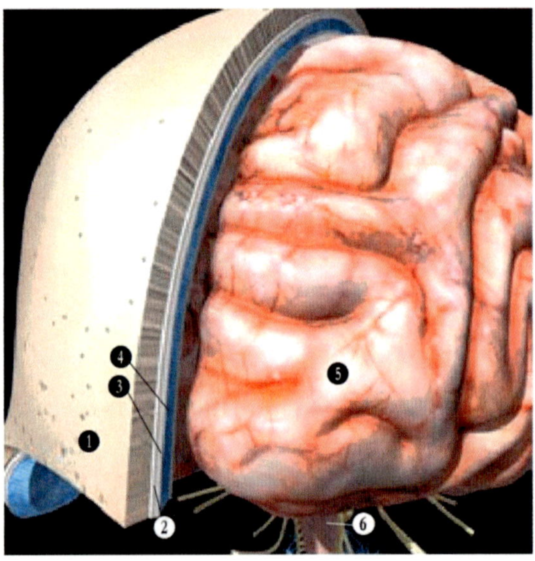

info

1. The word "meninges" is the plural form of the word "meninx", which is the Ancient Greek for "membrane". In some cases, these membranes become the target of microbial pathogens that cause inflammation, called **meningitis.** Unfortunately, this is a rather common problem in young animals, especially piglets. Because of the close proximity of the meninges to the brain, meningitis can easily progress to inflammation of the brain (**encephalitis**).

THE CENTRAL AND THE PERIPHERAL nervous system consist of two types of cells, the **nervous** cells, also called **neurons**, and the **glial**[1] **cells.** The human brain consists of approximately 100 billion neurons, and ten times as many glial cells! These numbers seem enormous and are certainly indicative of the complexity of the human brain, which is much larger than that of most domestic animals. Of course we must not forget the elephant whose brain contains approximately three times more neurons than humans, and weighs almost four times as much.

1. The word "glia" comes from a very old word of Ancient Greek meaning "glue", since these cells make up a glue-like substance in the brain. What may seem strange is that this term was first used in 1843 by a German medical doctor named Rudolf Carl Virchow, who obviously loved to study Ancient Greek!

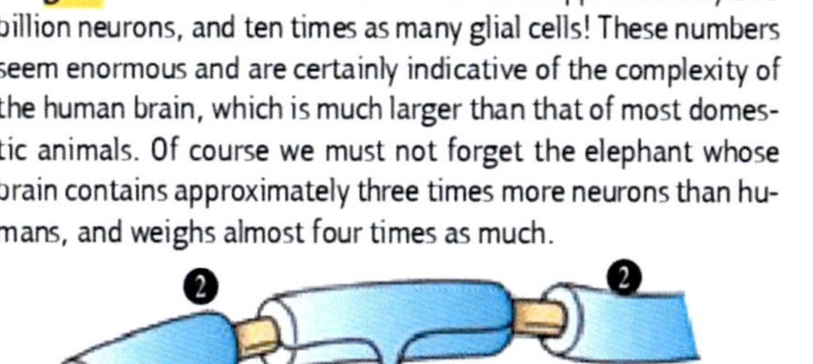

1 Glial cell
2 Myelin sheath

The size of the brain does not tell us much about how smart an animal species is. Another body measurement that could be used would be to assess the size of the brain relative to body mass i.e. how big an animal's brain is compared to the size of their body. Based on this criterion, the animal with the largest brain compared to its body mass is not the elephant or the human, but the marmoset monkey, a small mammal, not more than 20 centimeters in height that is native to South America.

Marmoset

The elephant has the smallest brain compared to body mass

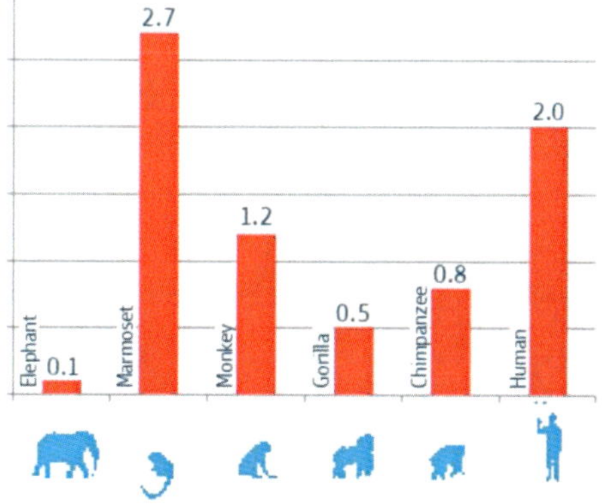

1. Soma, from the Greek word for "body"
2. Dendrite, from the Greek word "dendro", which means "tree"
3. There are some nervous cells located in the eye and the nose that have no axon
4. The famous English medical doctor Sir Charles Scott Sherrington who won the Nobel Prize for his studies in 1897, coined the term "synapse", which in Greek means "the point of contact"

Neurons are not what could be described as a typical cell. Of course, they have a cell body that is called **soma**[1], but they also have extensions; each extension is referred to as a **process**. There are two types of processes, the **axon** and the **dendrite**[2].

The axon looks like a stick, and can be short or long. This process of the nervous cell has the same function as that of the wire of an electric cable, connecting the neuron with its "target", such as an artery, a vein, an organ, a muscle or another neuron.

Dendrites are usually shorter than axons and are responsible for receiving most of the input from other neurons. Dendrites present several projections that make them look like branches of a tree, which is what their name denotes.

Nervous cells typically have one axon, and usually several dendrites. However there are some nervous cells that have no axon[3], and others that have no dendrites.

A stimulus generates a **nerve impulse**, which is a form of electric current that travels along the axons and is transferred from one neuron to the next, at specific sites that are called **synapses**[4].

As mentioned above, axons are surrounded by a lipid-rich substance (**myelin**) that is produced by the glial cells. Myelin forms a cover that surrounds the axon of the nervous cell and is called a **myelin sheath**.

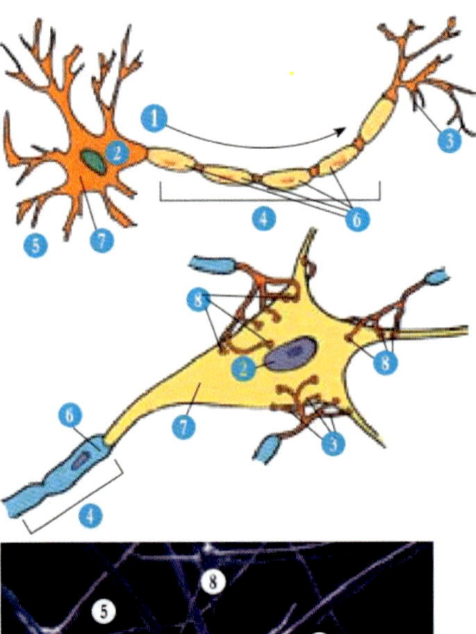

1 Direction of the nerve impulse
2 Nucleus
3 Axon terminals
4 Axon
5 Dendrites
6 Myelin sheath
7 Soma
8 Synapses

The myelin sheath is used to insulate the axon very much like the plastic cover that surrounds the wire of an electric cable. Obviously no electric cable can be used safely without its plastic cover, and no nervous cell can function properly without its glial cells and myelin sheath[1].
The axon together with its cover constitutes what we call a **nerve fiber**. A bundle of nerve fibers constitute a **nerve** (when located in the peripheral nervous system) or a **nerve tract** (when located in the central nervous system).

1 Nerve or Tract
2 Nerve fibber
3 Myelin sheath
4 Vessels (artery and vein)

1 Spinal cord
2 White matter
3 Gray matter
4 Spinal nerve
5 Meninges
6 Vertebra

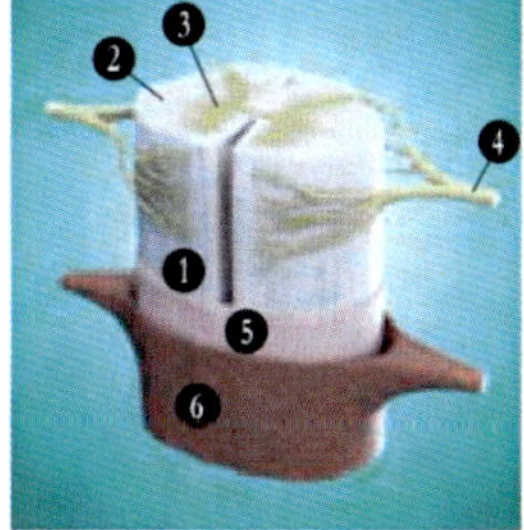

There are regions of the Nervous System that consist mainly of cell bodies and their dendrites. These regions have a pinkish or grey colour and are called the **grey matter**. These parts of the brain are often distinct from those that consist mainly of axons. Those regions of the Nervous System with many axons are referred to as **white matter**, because the myelin sheath that surrounds axons has the same colour as "fat" (white to bright yellow).

Based on the direction of the nerve impulse, neurons are divided into two categories, the **sensory** and the **motor**.
The sensory neurons are used for the transportation of nerve impulses from receptors to the central nervous system, whereas motor neurons transfer nerve impulses from the central nervous system to muscles or glands.

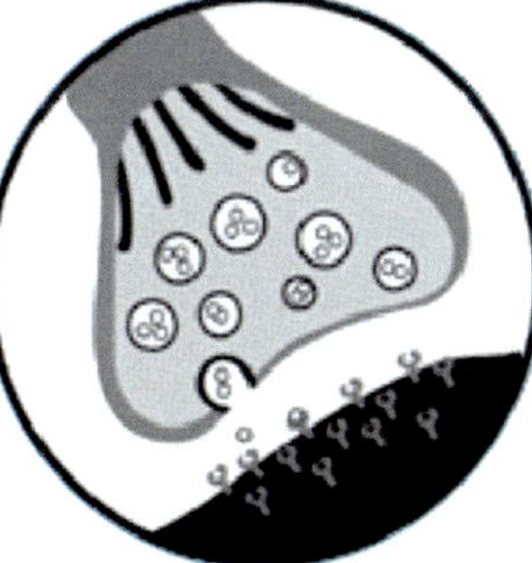

*Nerve impulses are transferred across synapses thanks to chemical substances (**neurotransmitters**) that travel from one nerve to the other, transferring the signal across the small gap that exists between them*

info

1. The myelin sheath can be destroyed by T-lymphocytes that for unknown reasons react to myelin as if it were foreign to the body. In humans, this **autoimmune reaction** (reaction of the immune system targeting the body's own cells) leads to a very serious neurological disorder, which is called **multiple sclerosis**. The same condition can also result from the dysfunction of glial cells. Disease process involving glial cells are also associated with **Alzheimer's** and **Parkinson's disease**. Although several treatments and medications can alleviate the symptoms of these diseases, unfortunately they continue to be incurable conditions.

Sensory function

The 5 basic senses and the respective sensory organs

Sense	Sensory organ
Sight	Eye
Hearing	Ear
Taste	Tongue
Smell	Nose
Touch	Skin

In everyday life, when we refer to the "sensory function" we tend to think only of the five basic senses of the body. However these senses refer mostly to input that the animal receives from its surrounding. As mentioned above, the Nervous System deals also with stimuli generated inside the body, such as for example the stretch of a muscle or the fracture of a bone. **In summary**, the mission of the sensory function is the collection of all the information that the animal needs in order to maintain consciousness of its body and surrounding. In healthy animals the flow of this information never stops and continues even during sleep, which means that the sensory function is constantly vigilant.

At this point it would be interesting to take a closer look at the sensory function in connection with senses that are very well developed in animals.

A cat marking a very good friend of hers!

THE SENSE OF SMELL is much more developed in most domestic animals compared to humans. In the dog, the part of the brain that deals with the sense of smell is much bigger than in humans. In contrast there are animals, like many marine mammals, that lack completely the sense of smell.

Animals use smell as a means of communication. In some species of deer, glands[1] located in the skin between their hind limbs release a smell that can be used by other deer to investigate if the specific individual is friendly or hostile. This makes it much easier to understand why when deer first meet they smell each other at certain regions of their body. Of course this type of behavior is very common in dogs that seem to find great joy in smelling each other's anal region ("the bottom"), were there is a gland with similar function. Cats have glands of this type in the skin of their forehead, and they use the smelly substance that these glands produce to scent mark humans or other animals and objects, by rubbing their head against them!

info

1. Gland, from the Latin *"glans"* meaning "acorn". The gland is usually an organ or a group of specialized cells (often shaped like the acorn) that produce and secrete a specific substance. One of the best known glands is the mammary gland of the female that produces milk and is a defining characteristic of mammals.

In some animal species, for example the pig, the sheep and the goat the smell of the male influences greatly the sexual behavior of the female, and vice versa. This is discussed in more detail in the last chapter of this book.

THE SENSE OF SIGHT is not the same in different animal species, and is adapted to their needs. This can be easily understood, if we observe the differences in the anatomy of the eye between **nocturnal animals**[1], i.e. those that are active mostly during the night, such as the cat, and **diurnal animals**[2], such as the dog. In nocturnal animals, the part of the external layer of the eye that is transparent to light is much larger compared to diurnal animals; this is one of the reasons that nocturnal animals have a very clear picture of their environment with much less light. Another reason is the structure of the inner part of the eye. In nocturnal animals the internal layer of the eye hosts a much larger number of receptors that respond specifically to dim light. These receptors are called **rods** because of their shape, whereas the receptors responsible for daylight vision are called **cones** (conical shape). The external part of the eye being more transparent, and the internal part carrying more rods, makes the eyes of nocturnal animals appear to have an unusual shine or to "glow" in the dark, when their **pupil** i.e. the hole in the centre of the eye that is surrounded by the **iris**, is **dilated** (is open wide) due to light being reflected off a special region at the back of the eye. The contrary happens during the day when the width of the pupils is decreased (**constricted**) to only a thin line, in order to protect the eye from being exposed to too much light.

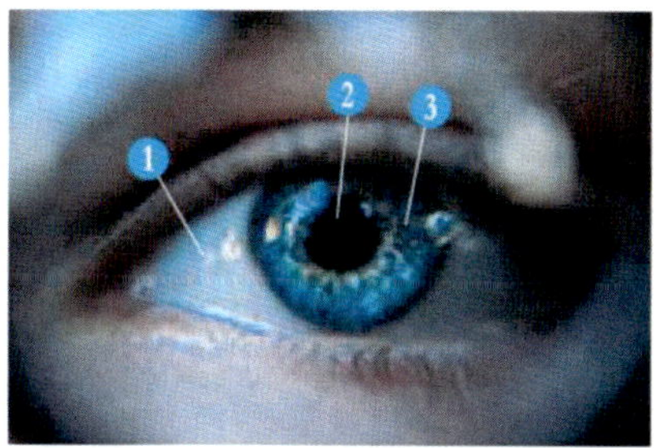

1 Sclera 2 Pupil 3 Iris

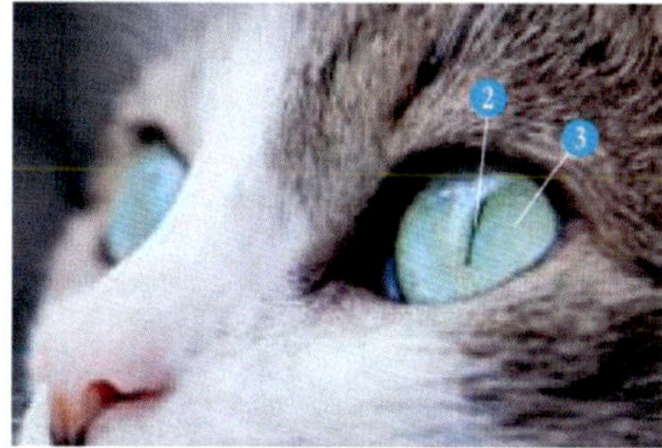

*Observe an interesting difference between feline and human eyes: the front part of the human eye around the iris is covered by a non-transparent membrane (**sclera**, meaning "tough"), which prevents light from entering. This membrane is hardly visible in feline eyes, which allows more light into the eye bulb and improves night vision.*

info

1. Nocturnal, from the Latin "*nocturnus*", meaning "belonging to night"
2. Diurnal, from the Latin "*dies*", meaning "day" + -urnus, which denotes time; hence diurnal means "that which is daily"

Another example which demonstrates how sight is adapted to the needs of survival is the different position of the eyes in predator and prey animals. Most prey animals have sideways-facing eyes, as opposed to predators whose eyes are facing forwards. Furthermore, when prey animals drop their head towards the ground to graze, their pupil rotates by up to 50 degrees to stay horizontal. This way, prey animals gain a wide width of vision, which means that they can see something approaching even almost from behind. However they lose the ability to discern the sense of depth, which has become limited to only a small part of their field of vision. In other words, these animals have a very good perception of all objects that surround them but not of their exact position in space. In contrast the perception of depth is very useful for most predators because they need to be fully aware of the exact position of their prey in order to make a successful pounce. Good perception of depth is a characteristic of **binocular vision**.

Even without binoculars, dogs have very good binocular vision!

Animals with developed binocular vision have forward-facing eyes, which creates an overlap between the images sent to the brain by each eye, and provides a narrow but very accurate three-dimensional perception of space.

Binocular vision is very well developed in predator birds, such as the eagle. In contrast the horse, like most herbivores, relies more on **monocular vision**, which is the result of having one eye located on each side of the skull. In effect the horse has a very broad peripheral vision, but doesn't see in the very front of its head, which is what we refer to as a **blind spot**.

Eyes located on each side of a very long skull is ideal for a broad field of vision, but impairs the perception of depth

The blind spot of the horses is a cone-shaped area that extends approximately 1 meter in front of their head. The only way that a horse can turn this blind spot into its field of vision is to move its head from one side to the other. Therefore when the animal is not able to turn its head freely, which happens often during horse-riding, the obstacles simply disappear from its vision when it gets close to them. This means that when a horse has to jump, it will have to measure the jump and determine its take-off point on the approach. When the obstacle is about 1 meter away, the horse will lose sight of it, and will have to rely solely on the rider for guidance, which of course requires a lot of experience from the rider and a lot of trust from the horse in the rider's abilities.

Jumps like this are amazing if you think that the horse is actually jumping over an obstacle that it can't really see!

HEARING IS ANOTHER SENSE that is very developed in certain animals such as the dog, cat, horse, elephant and rat. The receptors for hearing are located in the ear together with the structures responsible for the collection of stimuli necessary to maintain balance i.e. the vestibular apparatus, which is made up of three semicircular canals.

The ear is divided in three parts, the **external**, **middle,** and **inner ear**. Both types of receptors, i.e. those associated with hearing (**cochlea**[1]) and those responsible for balance, are located in the inner ear. The external ear consists of the outer visible part, which is called the **pinna** and the **ear canal.** The ear canal extends from pinna into the skull to the middle ear, and reaches a cavity that is called **tympanic cavity**[2]. In animals, the pinna is controlled by several muscles that make it very mobile and able to be directed towards the source of each sound. The sound waves are subsequently directed from the pinna through the ear canal to the middle ear, where they set in motion a membrane (**tympanic membrane** or eardrum) and a group of three small bones called the **hammer**, **anvil**, and **stirrup** because of their shape. This converts sound waves into mechanical action that is transmitted to the cochlea, and then to the brain, giving rise to the sense of hearing.

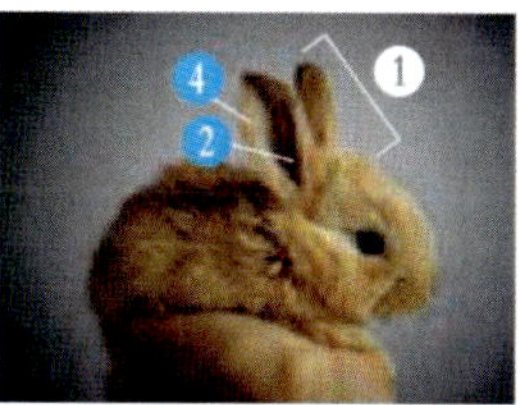

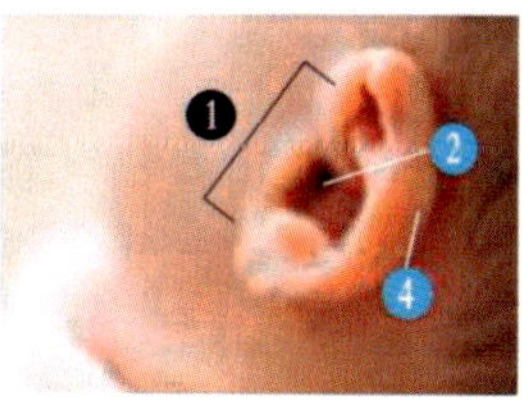

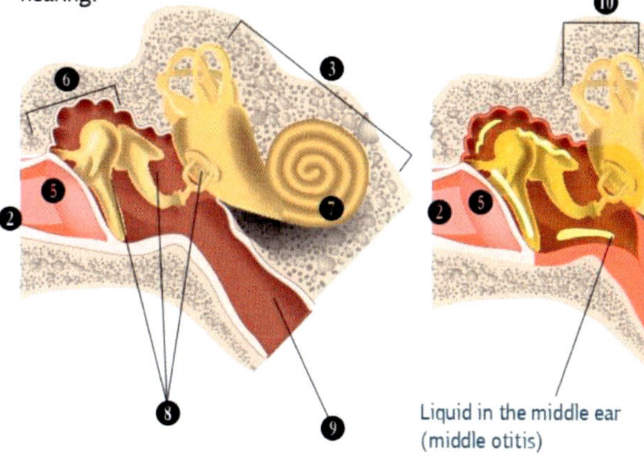

Liquid in the middle ear (middle otitis)

*10 A fluid (**endolymph**) inside these three semicircular canals is set in motion by head movement: head from left to right moves endolymph in the horizontal canal; nodding head movement activates endolymph of the superior canal; bending the head from shoulder to shoulder, does the same in the posterior canal. Endolymph movement activates sensors that transfer this "information" to the brain to form our sense of balance, acceleration and moving in a three dimensional space (**proprioception**). According to theoretical physics there are at least 10 dimensions of space, so how come we can perceive only three?*

1. External ear
2. Ear canal
3. Inner ear
4. Pinna
5. Tympanic membrane
6. Middle ear
7. Cochlea
8. Hammer, anvil and stirrup
9. Eustachian tube

1. Spiral shaped cavity; the word "cochlea" derives from the Ancient Greek for snail ("kokhlias"). Indeed cochlea looks very much like a snail.
2. The word "tympanic" means that which looks like a "drum" ("tympano" the Greek for "drum")

The range of frequencies that can be perceived by the ear varies among species. In humans the lowest frequency audible is around 20 cycles per second and the highest 20,000. The dog can hear sounds whose frequency reaches up to 50,000 cycles per second, which explains the use of dog whistles that produce high pitch sounds audible only to dogs.

Moths have the keenest hearing in the Animal Kingdom and this is an excellent example of evolutionary adaptation to increase the chances of survival.

Moth

But why do moths need such good hearing?

Bats are the main predators of moth. Bats also need a highly developed sense of hearing to survive, since they are practically blind and navigate during flight using **echolocation**, i.e. the perception of space generated by the reflection of the ultrasounds[1] they produce on the surrounding surfaces. Moths being able to hear sounds of higher frequency than bats, can evade their winged predators because they can actually "hear them comming"!

Integration

In order to understand the function of integration, one needs to remember that the nerve impulses that are transferred along nerves are essentially a form of electrical current. This means **that at this stage they cause stimulation but do not "contain" any information**. Before these electrical impulses are translated into information, they must be processed in specific regions of the Nervous System, i.e. the integration areas. When this happens, the Nervous System perceives the information that was transferred through the nerve impulse and reacts accordingly. **This two stage process, which consists of the conversion of the nerve impulse into information and the sub-**

info

1. Ultrasounds, i.e. sounds whose frequency is too high for the human ear (above a human's sonic range). The bat can hear sounds the frequency of which reaches up to 300,000 cycles per second.

sequent reaction, is referred to as integration.
Try to understand integration with the following example:

A dog approaches a bone. There are chemical substances released from the bone into the air and reach the dog's nose. These chemical substances will stimulate receptors located inside the dog's nose, which will generate a nerve impulse that will be transferred along a nerve (**olfactory**[1] **nerve**). As explained above, at this stage the nerve impulse is still only a form of electrical current and does not really provide any information about the smell or its meaning. The nerve impulses transferred by the olfactory nerve are directed to certain regions of the central nervous system where they are processed and integrated, i.e. converted from a nerve impulse into **what is perceived as a "smell"**. As you may imagine this information will be combined with "memories" and "experiences" stored in other parts of the brain, and together they will create the consciousness of the smell i.e. they will make the dog understand that the stimulus corresponds to a smell, and this smell belongs to a bone. It is almost certain that this integration will also generate emotions, such as hunger, restlessness and irritation, and it may even be combined with a very clear mental image of a bone, making the smell simply irresistible!

When threatened, the skunk sprays its opponents with a smelly substance that is produced by glands (anal glands) located in its anus (bottom). This liquid can even cause temporary blindness!

Motor function

The motor function is that which is responsible for the body's ability to respond to stimuli. Based on whether the response to a stimulus is voluntary or not, the Nervous System is divided mainly into two parts, the **somatic nervous system** which controls voluntary responses, and the **autonomous nervous system**, which controls all the rest. In more simple terms, if an animal made a conscious decision to run, this would require the activation of several skeletal muscles by the somatic nervous system. However it would not be possible to survive if similar conscious decisions were required in order to maintain the function of the heart, lungs or the intestines.

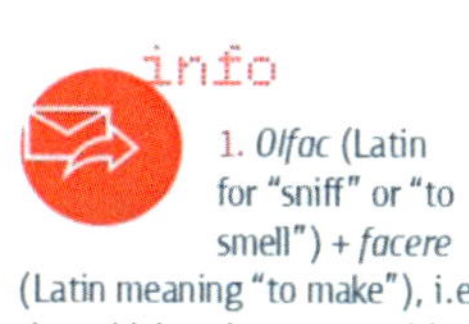

info

1. *Olfac* (Latin for "sniff" or "to smell") + *facere* (Latin meaning "to make"), i.e. that which makes you sensitive (be able) to smell

The activity of these organs is independent of the animal's will, and is controlled by the autonomous nervous system.

It has to be noted that part of the activity of the somatic nervous system is similar to that of the autonomous, in the sense that it is not entirely controlled by will. This is true in the case of **reflexes**. A reflex is an automatic contraction of skeletal muscles, that does not require a conscious decision.
An example of a reflex that is often tested in humans, dogs and cats is the **knee-jerk reflex** (or **patellar reflex**) i.e. the sudden kicking of the lower leg in response to a sharp tap on the front of the knee (patellar ligament), just below the kneecap.

Reflexes are generally activated when an immediate response is required, such as when there is a need to maintain body-balance or to withdraw one's hand from a hot surface.

Another example that probably demonstrates more clearly the purpose of these automatic responses is a reflex of dogs that is tested in animals being assessed for pruritis i.e. itchiness[1]: the veterinarian folds the ear (the pinna) in two and rubs one side against the other; if the dog is itchy, usually one leg becomes unstable, trembles and very often starts scratching intensively the neck and the head.

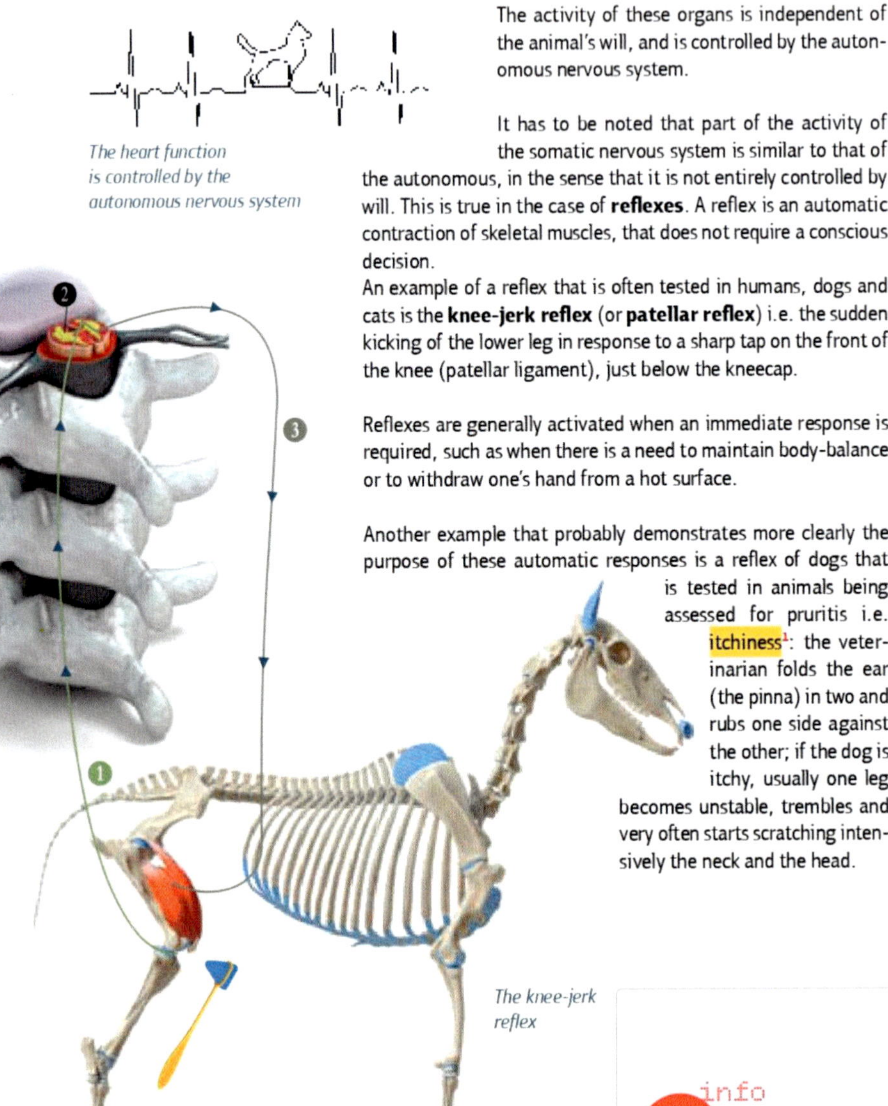

The heart function is controlled by the autonomous nervous system

The knee-jerk reflex

info

1. A dog will often not display any signs of itch when examined in a veterinary clinic, due to fear

Though activated through the somatic nervous system, this reaction is not really controlled by the animal's conscious will and aims to immediately remove the cause of the itch, such as ticks or fleas. In some cases vets use a rather funny construction to prevent self-inflicted trauma in dogs and cats due to intense scratching. This device is called an **Elizabethian collar** and it can be very helpful in dealing with this difficult problem.

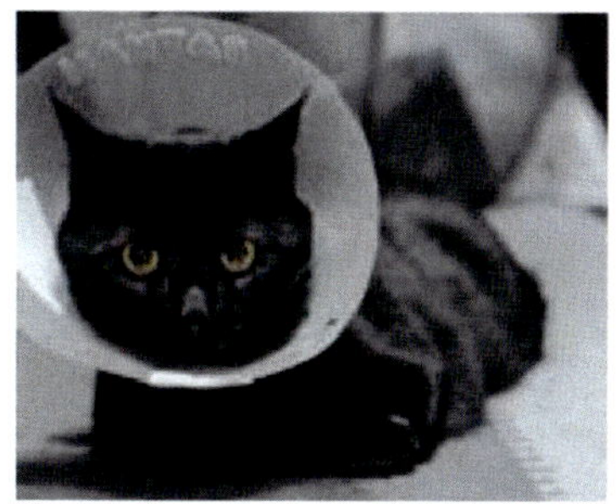

Of course don't ever expect to see an animal very happy at having to put up with an Elizabethian collar!

As you know, when animals are in pain they may become aggressive. Obviously this is a natural reaction caused by fear, which is generated because of the pain. These animals have to be treated with understanding and of course caution. Therefore when an animal requires an operation, it is necessary to find a way to decrease, or if possible to stop the sense of pain. In most cases this is done with **anesthetic drugs**, i.e. chemical substances that function mainly by inhibiting the activity of neurotransmitters, thus decreasing the animal's ability to feel and respond to pain. In this way, an animal can be submitted to a long operation without feeling any pain. There are, however, other ways to control pain in animals. A construction that is often used for this purpose is the twitch, which exerts pressure on small areas of the skin somewhat similar to a pair of pliers.

I used a twitch once, when I had to remove a nail from a horse's hoof. Horses are of course very strong, and when in pain can become dangerous, so I did the following: I asked my assistant to use the twitch to apply pressure to the upper lip of the horse's mouth. This immobilizes the horse and destructs its attention from other parts of its body that may be in pain, giving you just enough time for a short operation. Of course everything has to be done very carefully and quickly, because if the horse gets loose, chances are it will not be very happy to see you around!

CARDIOVASCULAR SYSTEM

The purpose of the Cardiovascular System of animals and humans is transportation around the body of chemical substances necessary for the nourishment of all body cells, and the removal of the chemical waste from metabolism[1]. In this respect the Cardiovascular System is of critical significance for the body's **homeostasis**[2].

The means of transportation of nutrients and waste products to and from body cells is **blood**[3], which circulates inside a vast network of tubes, called **blood vessels**. There are two types of blood in the body, **arterial** and **venous**. The former (arterial blood) carries nutrients, hormones and oxygen to cells, whereas the latter (venous blood) is used for the transportation of their waste products. In addition to arterial and venous blood, there is one more liquid tissue in the animal body, the **lymph**[4], which is essentially the fluid that surrounds the cells **(extracellular fluid)**.

This extracellular fluid which nourishes cells and absorbs their waste products is collected and removed from the **intercellular space** through a network of vessels that are called **lymph vessels**. This means that just like arterial blood circulates inside **arteries**, and venous blood inside **veins**, lymph is transported inside lymph vessels. One more very significant function of the lymph is that it provides protection against microbial pathogens, since it contains the **lymphocytes**, a type of cells specifically trained to recognize and kill invading microbes.

info

1. Metabolism is the series of the chemical reactions that take place inside a living cell. The waste products of metabolism, such as for example carbon dioxide, have to be removed from the body, because if they accumulate they become toxic, i.e. they can poison cells.
2. Homeostasis is the ability of an organism or a cell to maintain its physiological balance: "homeo-" the Greek for "similar to", and "-stasis" which means "standing still"
3. Blood is a tissue i.e. a group of similar cells with a specific function.
4. **From nymph to lymph**: in Ancient Greek mythology "nymphs" were goddesses of lakes and springs. In Latin, the word "nymph" became *"lympha"*, which was gradually used to denote "clear/colourless fluid or water". The word *lympha* is the origin of the term "lymph", since in contrast to blood, lymph is indeed colourless (clear).

Arterial blood is rich in oxygen and has a bright red colour, whereas that of veins is darker, almost purple. The arteries are under high pressure from the pumping of the heart, and to sustain this pressure they have thick, elastic walls. The veins are under much less pressure, and therefore have thinner walls. The circulation of blood between arteries and veins is mediated by smaller vessels called **capillaries**.

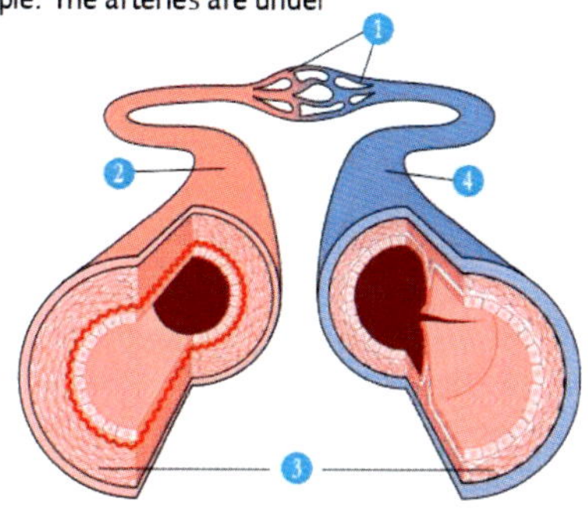

1 Capillaries
2 Artery
3 Layer of smooth muscles
4 Vein

A large part of the wall of blood vessels consists of smooth muscle cells. The smooth muscles of arteries and veins can contract, which decreases the width of the blood vessel lumen (i.e. their diameter) and blood flow. The exact opposite happens when these muscles relax: vascular lumen width and blood flow are increased. The width of blood vessels is controlled through nerve endings of the autonomous nervous system, so that blood flow is always exactly that which is necessary for the body to remain at its correct physiological condition, i.e. to retain homeostasis.

The core of the Cardiovascular System is the heart, the activity of which sets blood in continuous motion. As explained in the previous chapter, the heart consists of a special type of muscle (myocardium) and functions very much like a simple pump. This similarity is not limited to the function of the heart but extends to its structure:

The myocardium is covered with a thin membrane called the **pericardium**[1], which produces a fluid (**pericardial fluid**). The pericardial fluid fills the space between myocardium and pericardium, and has similar role to the lubricants that are used in all types of pumps i.e. to decrease friction and facilitate the continuous movement of the pump.

PET FIRST AID

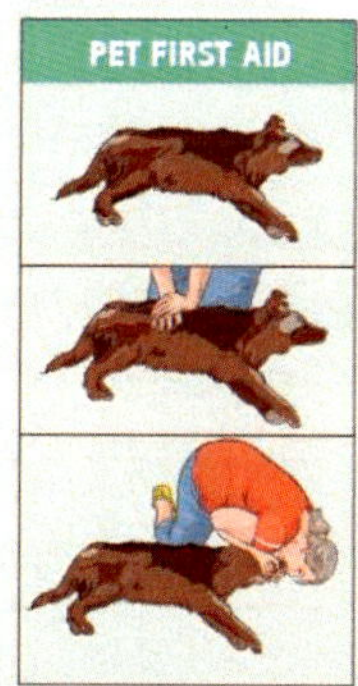

*When the heart stops (**cardiac arrest**), it is necessary to perform **CPR** i.e. **cardiopulmonary resuscitation**, by compressing the thorax and breathing air into the lungs, which in animals is usually done through the nostrils.*

info

1. A term of Greek origin; pericardium, that which surrounds ("peri-") the heart ("Kardia")

The volume of blood in average size adult animals

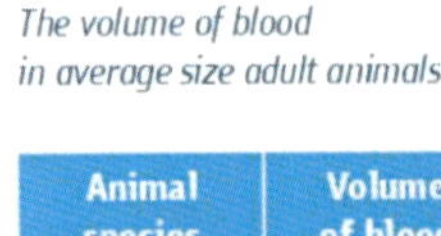

Animal species	Volume of blood (liters)
Elephant	250
Horse	55
Cow	40
Human	5
Dog	2.5
Cat	0.3

Auscultation with a stethoscope

In animals, the auscultation[1] of the heart (i.e. to hear the sounds of the heart with a stethoscope) is more difficult than in humans for many reasons. One of them is that the auscultation area is not as easily accessible because it is located underneath the animal's left elbow. This part of the limb has strong muscles, and in some animals it is pressed rather firmly against the chest leaving very little free space to access the chest wall.

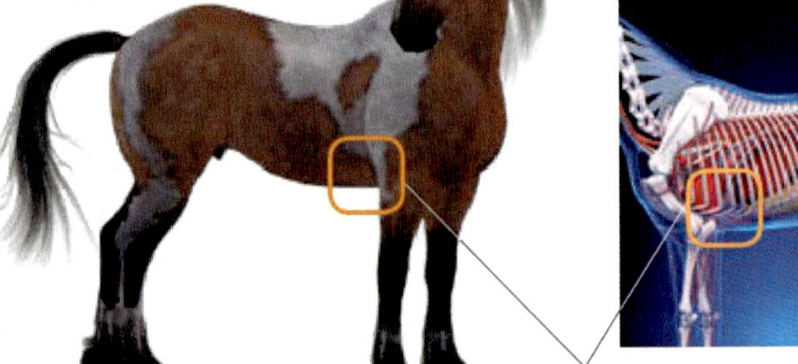

Auscultation area of the heart

The function of the Circulatory System

When the heart contracts, blood is pushed into the arteries with enough thrust (i.e. power) to reach rapidly even the most remote parts of the body.

The venous blood, which is poor in oxygen and rich in carbon dioxide, returns to the heart from all body parts and from there, is pumped to the lungs, to be **oxygenated**, i.e. mixed with oxygen. This means that two categories of blood pass through the heart, arterial blood that is rich in oxygen and is directed to all body cells, and venous blood that transports the carbon dioxide from the cells to the lungs.

Through the function of breathing, the carbon dioxide of the venous blood that has reached the lungs will be transferred into the air and will be replaced with fresh oxygen. The oxygenated blood

info

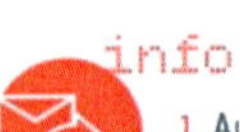

1. Auscultation, a term of Latin origin meaning "the act of listening". The word "stethoscope" is of Greek origin ("stethos" meaning "chest", and "scope" meaning "to see, to examine").

will return to the heart to be pushed again to the cells, completing blood circulation, which is usually described as two interacting loops.

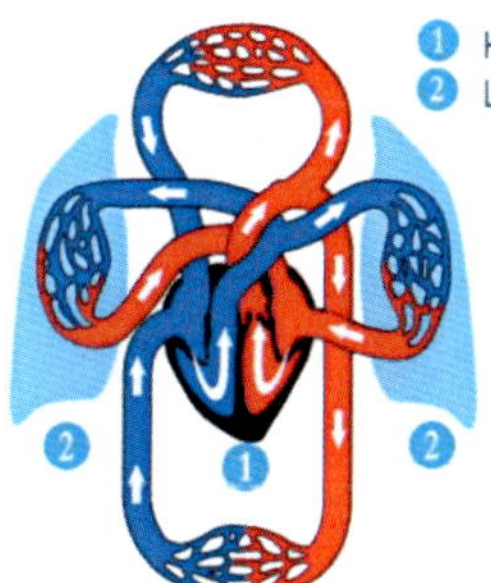

The two interacting loops of the circulation depicting oxygenated blood in red colour, and non-oxygenated in blue. Observe that the way heart is divided, oxygenated blood from its left side (red-coloured content) does not mix with non-oxygenated blood (blue-coloured content) that circulates through its right side.

The first of these loops corresponds to blood circulating between heart and body parts, and is called **systemic circulation**. The second loop corresponds to blood circulating to and from the lungs, and is called **pulmonary circulation**. The only part of the body were the two loops interact is inside the heart, which is divided into four compartments or chambers. The upper chambers are called **atria**[1] (right and left atrium), whereas the lower are called **ventricles**[2] (right and left ventricle). There is no connection between right and left atrium or between right and left ventricle. However the right atrium and right ventricle are connected through a valve, as are the left atrium and ventricle. The heart valves located between atria and ventricles allow blood flow only in one direction, i.e. from atria to ventricles.

Though mammals, birds and some reptiles, including the crocodile, have four-chambered hearts, the heart of many amphibians such as frogs and lizards have only three chambers (2 atria + 1 ventricle).

The heart of fish has only two compartments (1 atrium + 1 ventricle), which makes blood circulation less efficient and their life more difficult!

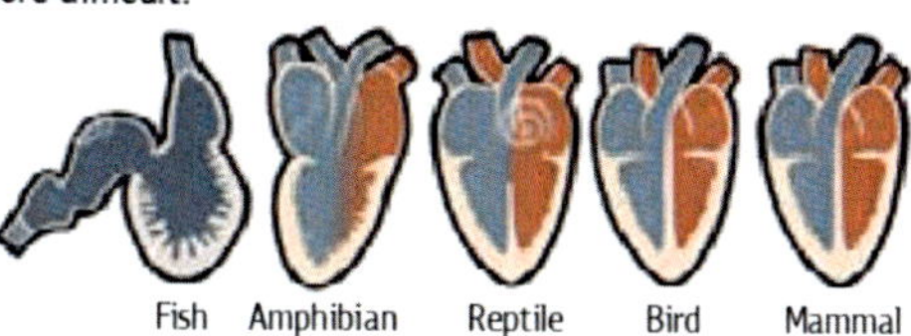

info

1. Atria, from the Latin *"atrium"*, which was the name of the room of the Ancient Roman house that had the fireplace (living room). The smoke escapes from the fireplace through the chimney, just like blood escapes from atria to ventricles.
2. Ventricles, from the Latin *"ventriculus"*, meaning "little belly"

If fish remain motionless for long they die!

In fish, the return of blood from the different parts of the body to the heart requires the pressure generated by their body being constantly in motion, since their heart cannot generate the necessary thrust on its own. This means that if fish remain motionless for long, they die because the blood pumped from their heart has insufficient pressure to retain physiological body activity (the pressure is just enough for sending blood to the body parts, but not for its return).

Strong body but weak heart!

The fact that fish need to be in constant motion to stay alive is possible only because they live in water, which facilitates their movement. Being in constant movement would be impossible for terrestrial animals (those that live predominantly or entirely on land), because it would require more energy than that they could possibly acquire through food.

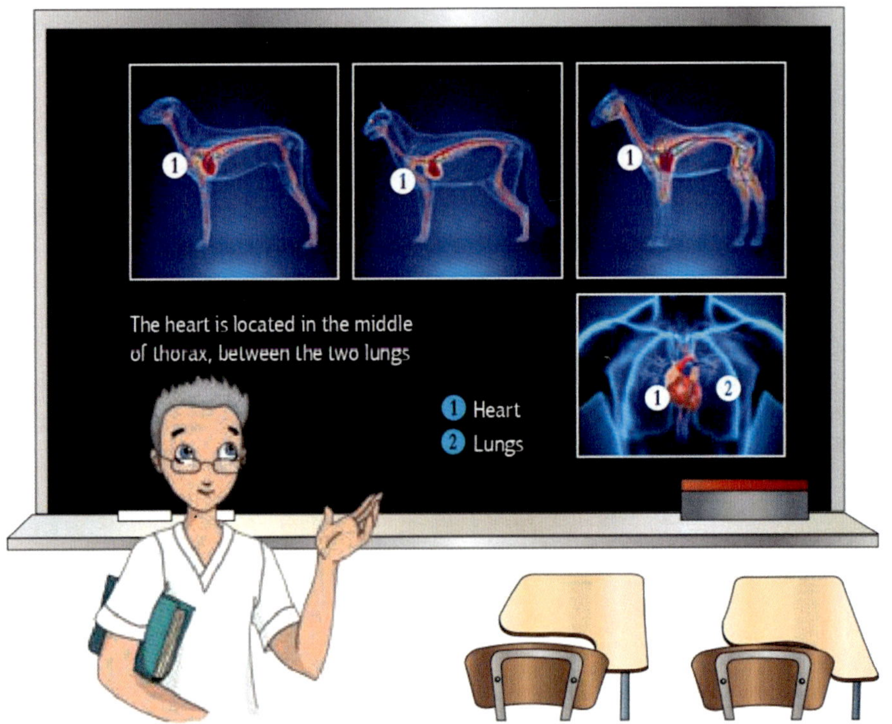

Stories from the life of Dr Hippocrates

There is a very serious disease that affects dogs, and less commonly cats and humans. This disease is called **dirofilariosis** (or dirofilariasis) and is caused by the parasite **dirofilaria**, which is commonly referred to as "heartworm", since it affects the heart. The disease became known many centuries ago and was feared very much. This is apparent even from the name of the parasite (from the Latin *"dirus"*, meaning "fearful", and *"filum"*, meaning "thread", i.e. the "fearful thread").

The fearful thread

The larvae termed microfilariae of the parasite circulate in the blood of affected dogs, and because of their microscopic size, they are passed easily to other animals and humans through mosquito bites. The larvae gradually turn into mature parasites that look like threads and can be very long. These "threads" form clumps that may get stuck inside the heart, the blood vessels and the lungs, causing great distress to the animal and possibly leading to death.

I came across my first case of dirofilariosis as a trainee in the final year of vet school. The dog, a male huntdog 6 years of age, was already very sick when brought to the clinic and could not be saved. During post mortem examination, i.e. that which is performed after death, we discovered that the heart of the animal was filled with these white thread-like parasites, some of which were almost 50 centimeters in length. There were so many of them that we had to use a bucket to collect them. I must confess that this was a shocking experience that still comes to mind, since I was overwhelmed by the thought of the animal having this number of worms inside its heart!

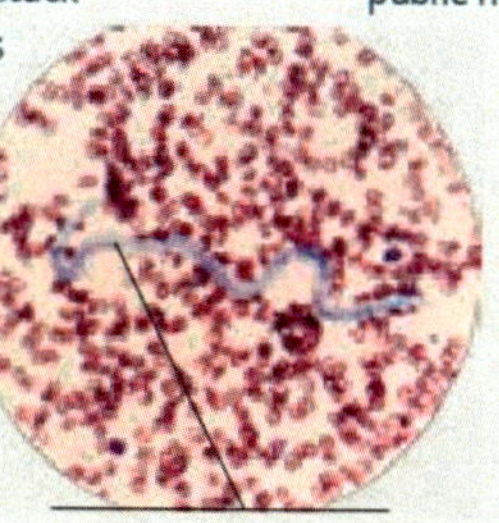

Microscopic view of dirofilaria in blood

Though it is still very difficult to treat animals with dirofilariosis, today there are several products that can be used in dogs as a means of protection, and can prevent the transmission of the parasite. These measures of prevention are recommended in all the parts of the world were mosquitoes live, not only for keeping dogs safe, but also for the protection of public health.

Made in the USA
Las Vegas, NV
21 August 2024

94246228R00031